50 DO-IT-YOURSELF PROJECTS FOR KEEPING GOATS

Fencing, Milking Stands, First Aid Kit, Play Structures, and More!

JANET GARMAN,
TIMBER CREEK FARM

FOREWORD BY MARISSA AMES,
EDITOR, *GOAT JOURNAL* MAGAZINE

Skyhorse Publishing

Skyhorse Publishing books may be purchased in bulk at special discounts for sales promotion, corporate gifts, fund-raising, or educational purposes. Special editions can also be created to specifications. For details, contact the Special Sales Department, Skyhorse Publishing, 307 West 36th Street, 11th Floor, New York, NY 10018 or info@skyhorsepublishing.com.

Skyhorse® and Skyhorse Publishing® are registered trademarks of Skyhorse Publishing, Inc.®, a Delaware corporation.

Visit our website at www.skyhorsepublishing.com.

10 9 8 7 6 5

Library of Congress Cataloging-in-Publication Data is available on file.

Cover design by Abigail Gehring
Cover image by Janet Garman

Print ISBN: 978-1-5107-5012-8
Ebook ISBN: 978-1-5107-5013-5

Printed in China

Contents

Foreword

"Get a goat," they said. "Goats are easy!"

Before you take that advice, ask your adviser if they have ever kept goats. Chances are, they haven't, and they're repeating what they've heard from other people who have never kept goats.

"Goats can eat anything," they say.

"Goats pretty much care for themselves," they say.

Before you buy that first goat, read up on what goats specifically need, because the truth is that they can die from eating too much of the healthiest goat food in the world.

Longtime goat owners will attest to how goats have specific fencing needs and feed requirements. They'll tell you how goats must have certain minerals, which may not be available in your local soil to transfer through forage. Or that a female nubian will scream all night as if she's in mortal danger, when all she wants is to meet a nice buck.

This isn't to discourage you from buying that goat (or goats, because they should never be kept alone), but to encourage you to educate yourself first. Or, if you've recently found yourself the first-time owner of a goat, educate yourself *fast*. Start with feed and shelter, then go from there.

Janet Garman, a longtime farmer and goat-owner, has written for internationally recognized agricultural magazines for years. With followers in the tens-of-thousands, she provides valuable information and advice through her Facebook page, blog, books, and the articles she writes for outside publications. In these pages, Janet goes over and beyond dairy goats to talk about fiber goats like pygoras, raising goats for meat, and even goat milk and cheese.

Do-it-yourself projects are a must for self-sufficiency, and Janet has a wealth of them to share. What do you do when a doe rejects her baby, and the feed store is a hundred miles away . . . and closed for the weekend? How can you prevent worms and handle bloat with no access to an on-call veterinarian? And do you have to pay someone to trim your goat's hooves, or can you do it yourself?

I recommend Janet's book to all new goat owners. Within a few chapters, you will feel more confident about keeping your goats healthy and will be excited to create projects such as goat milking stands, hay racks, feed supplements, and even goat playgrounds for health and mental enrichment.

Congratulations on goat ownership, and good luck with all things goat!

—Marissa Ames
Editor, *Goat Journal* magazine

Disclaimer

While working on the projects in this book, we made certain variations based on our stalls and available materials. If you possess a basic knowledge of how to build framed structures, that will be helpful, but I have tried to make the instructions clear enough that anyone can complete the projects, even with no prior building experience. I have made every effort to give clear, detailed instructions, but we also believe in using what we have on hand rather than buying new materials for projects, and we encourage you to do the same. Look around your property. Do you have building materials that you can repurpose? Make do and reuse, but keep in mind that using different materials than are specified in the instructions may affect the final measurements. Please remember to always wear safety eye protection, gloves, and dust masks when appropriate.

Kinder goats. Photo by Dillon Irwin.

No advice or idea stated in this work is meant to take the place of a licensed veterinarian caring for your goats. Before holistically treating any illness or injury, be aware that the supplemental care and first aid described in this book is in no way intended to take the place of veterinarian care.

HOMER'S
ALL-PURPOSE
BUCKET
5 Gallons

Introduction

Goats were among the first animals that humans domesticated. For over ten thousand years, goats have contributed to rural living and provided specialty and luxury products to the general population. Goats provide meat, milk, labor, and fiber. They are excellent companion animals, too, and add a certain element of fun to any backyard farm. Although they are considered livestock (and if you have more than a few, that's how you will run your operation), they are also a lot like outdoor pets.

It's hard to name another animal species that provides as much as goats can . . . all while keeping us guessing about the next adventure they have planned. You may end the goat-owning journey feeling as baffled by goats as you were on day one. They are mischievous and capricious creatures, but chances are, you will quickly start to file away important and practical caretaking info and methods as you live and learn together with your goats. There will be times you feel like throwing in the towel. Goats do that to a person. I encourage you at those points to keep going. Those disappointments will build your skills and knowledge if you let them.

Keeping goats housed, fed, and healthy can be done in quite a few creative ways! Goats are happy with simple structures, making them the ideal recipients of home built, do-it-yourself projects. They are just as content climbing on a half-buried tire as they are on a structure costing hundreds of dollars. I prefer to create from recycled materials and the scrap lumber pile, although I do come up with ideas while browsing the home improvement store, too.

At the other end of the spectrum are all the wonderful projects you can master *because* you have goats. From the food options in the meat and cheese categories, to the crafts you can create with the mohair fiber, to the creamy soap recipes you can try, goats have much to offer do-it-yourselfers.

Please keep in mind that goats are rough and destructive. I knew this, but inconveniently forgot it when starting a few of the projects for the book. Anything that the goats have access to should be built to withstand constant assaults, body slams, and head butts. Don't worry, though. The destruction was a good reminder and all the projects in the book are designed with durability in mind.

What will you need in order to complete the projects in this book? A vision of what can be done with repurposed items will help. If you don't have that gift, the step-by-step instructions will assist you. Grab a few common household tools—a hammer and a rechargeable drill driver or a pack of screwdrivers are a good start—and dust off your basic building skills and kitchen know-how. We are going on an adventure to create useable items that enrich your goats' lives and make goat-keeping easier for you. From the kidding pen to the cheese plate, to gifts for family and friends, and products for your home, you'll find lots of ideas.

I hope my book encourages you in your journey with goats. Keeping a species that has supplied us with countless years of milk, meat, fiber, and work is humbling. They can teach us many good lessons about animal husbandry. Goats keep us laughing, sometimes make us cry, and never stop trying to win us over with their antics. Because of all they have added to our farm over the years, along with the many frustrations, I think they deserve to have the best environment our farm can provide. It doesn't cost much at all to complete the projects in this book. Care was taken to show possibilities for many different farm and homestead scenarios. Take a look at your scrap lumber pile and recycled building parts in a whole new way. Your goats will thank you.

Let's get started!

Preparations

Before You Buy Goats

Before signing the papers to purchase any farm animal, including goats, make sure you are in the proper zoning for owning goats. It's a sad day on the farm when you realize that you have to give up the goats because the neighbors are complaining to the zoning board. While you and I might have a hard time believing this, not everyone thinks owning farm animals, or living nearby those who do, is the best life. It's better to skip the heartache and follow the rules. Remember, if owning farm animals is your goal, check that possibility when searching for a homestead site or new larger home with acreage. Owning a large parcel of land doesn't automatically mean you can legally own livestock.

Getting Started with Goats

The first thing you will need when bringing home your first goat for the farm or homestead is . . . more than one goat! I never recommend that anyone buy more animals than they can comfortably handle and provide for. But in the case of herd animals, like goats, a single goat will be lonely. And unhappy. And will try to escape in order to find its herd.

Besides deciding how many goats to purchase, you should also think through your goals for goat ownership. Knowing what type of goat farming you want to try will help you answer the questions about fencing, housing, diet, supplies needed, and other preparations. Are you most interested in goats for their milk, meat, fiber, help controlling brush, or companionship?

Since goats have existed for centuries, there are many breeds and much mixing of breeds. If you're purchasing goats for land clearing and brush eating purposes, the breed may not be a big factor, as pretty much all goats will happily clear your property of unwanted growth. If that is your goal for the goats, purchasing wethers (neutered males) will be your best choice economically.

The following six broad categories of goats will help you decide what breed is right for your homestead.

Dairy Breeds

Goat milk far outpaces cow milk in the world's dairy supply. Statistically, up to 65 percent or more of the milk consumed worldwide comes from our caprine friends. In the United States, the percentage is lower, but it is rising. The five most popular dairy breeds are:

- Saanen
- Toggenburg
- Nubian
- Olberhasli
- Nigerian Dwarf

Keep in mind that the does will need to be dried off and then re-bred each year in order to continue the milk supply. This does not mean you need to keep a buck, but if you grow your goat farm to a significant size, it might be easier to keep the buck rather than borrow one for breeding season. Bucks bring their own challenges and needs. You can read more about this subject in Chapter 9.

Nubian/Saanen doe. Photo by Victoria Young.

Fiber Breeds

Fiber goats produce mohair, which can be processed into soft roving and yarn. Angora goats, Pygora, Nigora, and Cashmere goats are all considered fiber goats and are primarily kept for the twice-yearly harvest of fiber. The renewable resource can bring extra income to your homestead or farm because mohair and cashmere are considered luxury fibers.

Fiber goats require the same care and housing as any other type of goat. They require dry housing just as dairy and meat breeds, especially during wet or cold weather. Their coats keep them warm, but they don't shed water like a sheep fleece will. Goats will get wet, and then chilled, often leading to illness.

Angora goats produce mohair, which is often confused with the fact that angora rabbits produce angora fiber. The mohair of true angoras yields a soft, warm fiber sought by luxury markets.

Angora goat. Photo by Mandi Chamberlin.

Cashmere goats are a type and not a breed, although many breeders have perfected breeding for the cashmere quality in their goat herds.

Pygora goats. Combining the best qualities of both the pygmy breed hardiness and the Angora goat fiber and temperament produced a small- to medium-sized goat capable of producing a twice-yearly clip of soft luxury fiber. Due to the breed background, Pygora fiber contains some guard hairs from the pygmy genetics that are preferably removed during processing or by hand.

Pygora goat.

Meat Breeds

Meat breeds are gaining popularity in the United States. As a goat farmer, it's important to accept that all goat breeds can be meat goats. If you are breeding, unless you have unlimited pasture and forage on your land, the excess male goats will need to be sold, or butchered for your own food. Breeding of any of the goat types will soon lead to overcrowding if you don't have a plan in place to sell or process the offspring that don't fit your goat goals. A dairy breed of goat will still yield a tasty meat product.

Kinder goat. Photo by Dillon Irwin.

That said, the typical meat goat breeds produce muscle weight more economically than dairy breeds. The market for goat meat is good, and many ethnic populations seek quality goat meat for religious and cultural celebrations. Spend some time researching the goat meat demand and cultural market requirements for your area. Some require that the goats be a certain age, and neutered versus intact males, and the presence and absence of horns are also factors.

Boer. The boer goat breed originated in South Africa. Their high rate of gain and growth has made them a popular choice for goat meat in the United States and other countries. Their appearance is a stocky build, mostly white with brown accents. Boer goats are considered easy keepers for goat farmers and easily fit the needs of a smaller homestead operation, too.

Boer goat. Photo by Fripp Family Farm.

Kiko. This New Zealand breed is gaining popularity due to its hardiness and meat quality. The Kiko is a result of crossing Saanen or Nubian bucks and wild goats, producing a goat that withstands varied climates and resists health issues.

Spanish. The Spanish breed has been in the United States since the early settlers arrived. Many breeds of goats have some of the Spanish breed influence as people traveled and interbred Spanish goats with other goat breeds being brought into the country from other areas.

Tennessee Fainting Goats. John Tinsley, a traveling goat owner, spread Tennessee fainting goats throughout the southern part of the United States during the nineteenth century. The goats became very popular due to the excellent reproduction rates and excellent meat quality. While there is a myotonic breed association with strict criteria, fainting goats can have a widely varied appearance. Being a Landrace and adaptable, you will find this breed ranging from 50 pounds to 175 pounds!

Kinder. The Kinder breed was developed more recently and is an American breed begun from necessity on a family farm in Washington state. The Kinder is a true dual-purpose goat, producing quality milk and meat. The result of breeding Nubian does with a Pygmy goat buck, the Kinder gained popularity for small farms that required milk that had high fat content for cheese, and fast growth for meat. Kinders are prolific non-seasonal breeders. Most breedings result in triplets and twins. Add to that a calm and happy goat nature and medium size and you can see why the Kinder breed is growing in popularity.

Kinder goat. Photo by Dillon Irwin.

Brush/Land Clearing

Almost any breed of goat will be happy to help you clean up your yard of weeds and brush. Goats love to work this type of job and will reward you with a clean property. Caution should be used if you have limited fresh forage. Carefully check that the property does not contain potentially toxic plants in addition to the healthy forage. Goats will choose the best food but, when that is limited, they will eat the toxic plants too. Supplement with hay and remove the toxic plants the goats have access to.

Pack Goats

Any goat can be trained as a pack goat. The conformation will determine the success of each animal at carrying bags and supplies. Look for sturdy, well-built goats with a flatter back and strong legs. Training should start early but a true pack goat will not be ready to carry heavy weight for a few years. Conditioning your goat from an early age will train their bodies as well as their minds to the task. Goats won't be able to pack as much as mules or donkeys, but they can be a viable animal especially in rocky mountainous terrain.

The same principles would apply to cart training. For centuries, goats have been used for pulling a cart of farm produce and even farmers.

Pet

I've never met a goat that wouldn't want to claim pet status. Goats love to be spoiled and pampered. Caution with the larger breeds! They will still think they are lap goat–sized even when they are full grown.

Photo by Mindie Dittemore.

What Are Wattles?

Wattles are fleshy appendages that hang from the neck or throat area of a goat. Occasionally they are found by the ears. They really have no known purpose. Some breeders cut them off soon after birth. Any breed can have wattles but not all goats have these interesting accessories!

Wattles on goats.

Endangered or Rare Breeds

Choosing a goat breed is a challenge. With so many breeds available, even considering only the breeds in the United States, it's a daunting task. What if you need a goat that fulfills more than one purpose on your homestead?

Breeds on the conservancy list can be great choices for a new goat owner who is unsure of their reason for keeping goats. Many breeds, such as the Arapawa, are perfect for a homestead, bringing excellent temperament, hardiness, milk, and meat to the deal.

Arapawa goat. Photo by Michelle Nardozzi.

The livestock conservancy has gathered much information about known breeds of goats that are in danger of extinction or critical status. If any of these breeds are available in the area you live, this could be a great start for your goat keeping journey. You would also be helping a breed return to a recovering status.

The Arapawa. The Arapawa goat breed still lives on its original habitat, an island off New Zealand. Although it is a feral breed, it adapts well to domestication, has a steady, calm temperament, and can be kept for meat. The milk production has not been measured as a true dairy breed, but for a family's use it is more than adequate. These animals are beautiful and a nice medium size for homesteaders. Arapawa goats are larger than pygmy goats and smaller than the full-size Toggenburgs. Since they have survived hundreds of years, their foraging ability is strong, and they possess great hardiness against disease.

The Arapawa goats have survived hundreds of years on the island, but now the New Zealand government feels the time is right to have them removed. The goats are in the right place at the wrong time. Tourism and farming are the primary economic interests of the island. Although the goats are not aggressive, feral goats wandering around and eating crops is problematic for the island's current goals. Breed conservationists in the UK, USA, and New Zealand are working to get these native goats into safe breeding programs and preserve the breed. As a homesteading proponent and educator, I was interested in the breed. I reached out to a breeder in the United States for some firsthand knowledge.

Photo by Michelle Fraser.

Michelle Nardozzi from Newbury Farms has added Arapawa goats to her family's homestead. The goats produce milk for the family. (Goat milk is a wonderful alternative to cow milk for those with allergies, and for those of us who just prefer the creamy goodness of goat milk.) While the Arapawa isn't a dairy goat in the same way the commercial dairy breeds are bred for production, the amount of milk more than fulfills the family's needs.

In addition, Michelle is a breeder, working to increase the presence of the breed in the United States to keep the breed thriving. From a breed conservationist standpoint, it's important to focus on the future of the breed. Choosing the best buck for the family goat yard is an important part of this plan. Currently, Phil, the resident buck, is doing his part to create cute kids. But as any good breeder knows, eventu-

Arapawa goat. Photo by Michelle Nardozzi.

ally Michelle will have to acquire another buck to keep the genetic lines healthy. The Arapawa is the most genetically distinct of the goat breeds.

In addition to milk production, this breed is excellent at clearing brush and weeds on the homestead. As with most species, goat meat is an option, although the primary goals of the conservationist currently are to do careful breeding and increase the numbers. This breed kids easily with little to no assistance needed. That in itself is an excellent advantage for a new goat owner. The breeders are looking for people committed to furthering the return of the Arapawa breed. With approximately fifteen breeders available in the United States, this is an opportunity for the right farm to make a difference in the breed's future while enjoying a hardy, parasite resistant, easygoing, small-sized goat. They are social and fun goats, and Michelle also stated that there are no aggression issues with her buck. Now that's a great reason to further investigate this historic breed!

Of course, other critical status breeds exist and need conservation help. The San Clemente Island Goat is also a genetically distinct breed that went feral on an island off the coast of California. Although they look like a smaller cousin of the Spanish goat, they are not genetically similar at all. They are small but not a dwarf. They are hardy and considered a multi-use goat.

For more information, check for a breed association that can put you in contact with a reputable breeder. A friendly relationship with a conscientious breeder is valuable and many are willing to mentor you through the process, because they want the best for the breed.

How Many Goats Should I Start With?

One goat will not be a happy goat. Goats rely on herd mates to warn and relay danger signals. When they have no comfort from a herd, goats will feel vulnerable and stressed. Sometimes a lone goat will bond with a pony, horse, or sheep. This is, occasionally, a workable situation.

While each property and owner will have certain limits on how many goats are a good size herd, I recommend starting with three. Two wethers and a milking doe gives you a good start, provides milk, and should something happen to one goat, you still have two to keep each other company. If you want to continue getting milk, you can choose to borrow a buck for mating or take your doe to the buck's farm for a few weeks. Keep in mind that you will now be expecting one, two, or three additional goats to join your herd. Begin making arrangements or plans for keeping, selling, or somehow re-homing the weaned kids in the near future. That is, unless you are comfortable with a growing goat herd.

Why Does My Goat Faint?

Myotonic, or "fainting goats," do topple over for reasons that we often don't understand. The fainting response is genetic and is not limited to the myotonic goat breed but is seen most prevalently there.

When goats with this condition startle, their legs stiffen from a lengthy contraction of the leg muscles. But it's not a true faint; the goat remains conscious and tips over onto its side. Normally the goat recovers quickly.

Tennessee fainting goats are the most prominent breed displaying this fainting characteristic. Tennessee fainters got their start when goat owner John Tinsley was traveling through the United States from Nova Scotia and sold a few of his stiff legged goats to some farmers in Tennessee. These early recipients of Tinsley's goats found they had great reproduction rates, and good muscle and meat quality.

The same goats became known as "wooden leg goats" when they arrived in Texas. Other names include "stiffs," "nervous," "scare," and "Tennessee meat goat." The Myotonic Goat Registry states consistent traits that are very important and need to be preserved. A myotonic goat has these characteristics:

1. Possesses a docile temperament.
2. Myotonia congenita leading to stiffness and muscularity (the gene for myotonia congenital is also responsible for the excellent muscle quality).
3. An abundance of high-quality muscle.
4. Adapts well to low-input forage feeding system.
5. Genetic distance from other breeds such that crossbreeding yields great hybrid vigor.

What should you do if your goat faints? First, never provoke a goat into a collapse or faint. Observe the goat to make sure it only fainted and didn't choke or become injured. Fainting goats will recover quickly. If you notice a pattern that causes your goat to faint, consider changing your routine. We noticed with our fainter that food caused a good deal of excitement for him and the other goats. This led him to have an attack and faint. It looked very much like an epileptic seizure for our goat but can appear differently in others. Calm behavior and a few minutes of gentle reassurance brings him back around. Separating the goats for feeding helped cut down on the frenzy and the resulting fainting spells.

For more information on the Tennessee Fainting Goat breed, consult the Livestock Conservancy website (livestockconservancy.org). The breed is currently in recovery status after falling out of popularity for a few decades.

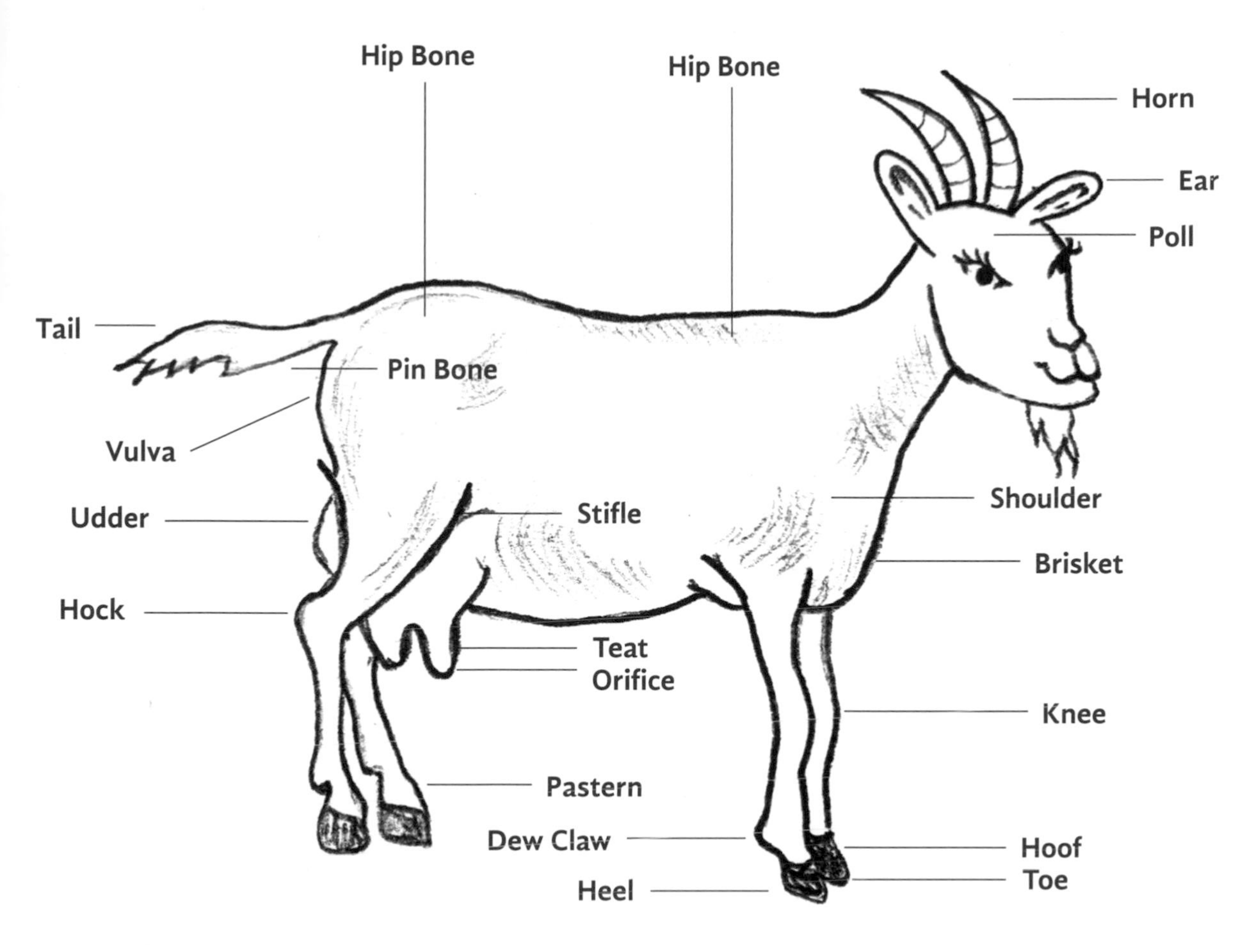

Body parts of a goat.

Can You Care for Goats?

Baby goats are cute and many even allow cuddles. What could possibly go wrong? While I will encourage you to educate yourself before you buy any animal, I do understand the impulse factor, too. I may have even succumbed to it a time or two.

Even though the kids start off sweet, gentle, and oh so adorable, eventually they will try to take advantage of you. With larger goat breeds, they may outweigh you before the first year is over. Bucks especially will transition from gentle kids who sit in your lap, to rambunctious, and maybe even aggressive when fully grown. This is how a buck would protect his flock in the wild and it's an instinctive behavior. Make sure

Continued on next page . . .

you take these factors into consideration when building your goat herd and choosing breeds.

Are you physically strong enough for goat herding? There are ways to handle large stock that do not require you to be stronger than the animal. Ask a goat keeping friend to let you help with her goats so you can familiarize yourself with their behavior. If you don't have any goat-keeping friends, ask a breeder that you are considering buying from.

Kiko goats. Photo Credit Denise Smith

The same principles apply to adopting goats from shelters. Take a hard look at all the information about the goat in question. Was she found wandering? She might be an escape artist. Is he aggressive? You may be adopting a dangerous animal. Was she turned in because she would not get along with the herd? She might have underlying health problems. While we all wish that animals in rescues would find homes, your home may not be the right one. And, while the price of a rescue animal might be enticing when you are building your flock, the adoptive animal can cost you more in the long run.

Your goat's daily care requirements are also a consideration. Will you be able to carry the food, hay, and water to the goat area? Is lifting a fifty-pound sack of goat grain possible, or do you have someone available to do that for you? These concerns might sound trivial to those who have raised livestock before, but they are important considerations for someone just breaking into small farming and animal husbandry.

Costs in both dollars and hours should be considered. Over the years, I have seen new farmers take on too much at one time. Adding large gardens, building infrastructure, raising a family, and working off the property at a daily job doesn't leave much time for animal care. Goats are not hard to keep, but they can be the straw that breaks the camel's back, especially if they escape regularly.

What exactly is required daily? You'll want to observe each animal carefully every day to make sure they are acting like healthy animals. Water should be emptied and refilled at least once a day, or twice a day during hot weather. Provide hay if the goats are not out foraging. In the winter, a small amount of goat grain should be fed to the herd. By "small amount" I mean a handful at most, that will keep them eager to come to you. Milking does and growing kids may require more grain.

Weekly, do a good scrubbing of buckets, and refreshing of stall bedding as needed. Pastures should be walked to check for any fence weakness or breaks, and to look for toxic plants growing up in the fields. Monthly hoof checks and trims should be performed as needed. Ours usually need a hoof trim every six to eight weeks, but it's good to check more frequently.

Another factor I have seen people overlook is the cost of keeping goats. At this point, you may have plenty of fresh growth on your property. But if there is a serious drought year, or some other devastating event such as a flood, are you financially prepared to purchase hay to meet the goat's forage needs? Take into consideration the cost of fence repairs, veterinary expenses, and first aid supplies, too.

Chapter 1: Shelters and Housing

Hopefully, before you reach the point of building your goats a home, you have researched the local ordinances regarding keeping goats and other livestock. Once you have that out of the way, it's time to move forward deciding what type of housing to build for your soon-to-arrive herd.

Your goats will demand a dry home to call their own. Goats can share a barn with other livestock, or you can provide a separate structure for goats. While goats are hardy animals, they do not like to be wet. Providing a safe, dry space for the goats to ruminate and sleep is important.

Goat housing can be a simple three-sided field run-in shed or a more complex goat barn. A stall in the barn with other livestock may be sufficient, too. One ideal setup is to have a shed large enough for all the goats to lie down in, as well as a couple of water buckets. Off of the shed, you could have a fenced paddock (make sure the fence is sturdy) with well-drained ground, and an overhanging roof for a dry hay feeding area.

Garden sheds are often converted into goat housing, and pallets can be used to fashion an inexpensive and yet sturdy goat shelter. Depending on your weather and climate, a three-sided, open-front shelter may be adequate, or you may need to provide housing that is more protected from the elements. In any event, housing for goats does not have to be elaborate or expensive.

If you plan to breed your goats—a distinct possibility if you are going to have milking goats—you will want the ability to section off part of the goat housing for kidding stalls or jugs. These areas provide some peace and privacy for the new doe during labor and delivery and while bonding with her new kids. Most goat breeders will move their pregnant does inside before the actual expected date of birth. This structure can be enclosed in an existing barn, or a shed that is renovated to include small stalls for the mamas and kids. A field shelter may even be adequate for your breeding stock if you can check on the goats frequently. It is not ideal, because the does may choose to give birth in the field, leaving the kid vulnerable to wet ground, cold temperatures, and predators. The best goat shelter for your breeding stock is an enclosed, well-ventilated, draft-free building.

A separate but nearby area can serve as a milking parlor. Having a milking area that is protected from the weather will make your task easier and more pleasant.

If you decide to keep a buck, you may want to keep him separate from the rest of the flock. Often people will house a wether with the buck for company. Keeping a buck with a doe in milk can sometimes cause the milk to taste off. Keeping the buck separately allows you to control the breeding dates and have a window of time that kidding will occur.

Stalls and shelters will need a soft, absorbent bedding for the floor of the stall or building. Straw is a great bedding material. The hollow core of the straw makes it a wonderful insulator. Also, when raising fiber breeds such as Angora or Pygora or for sheep, the straw won't burrow into the wool as much as sawdust or wood chips can do.

Depending on the type of goat you are raising, packaged bales of pellet bedding, shavings, or sawdust can be used. Make sure the bedding stays clean and dry. Hay is not the best choice for bedding. Hay has a much higher moisture content than straw and can become damp easily.

Even if your goats are weather hardy, providing shelter is one of the essentials of animal management. The goat shelter or barn does not have to be elaborate. The animals will appreciate the cozy home to rest in during the winter days and cold nights.

Sheds and Shelters

Pole Building

We built our field shelter originally for cattle. It is a pole shed that backs up to a natural embankment for wind block. The roof is made of corrugated tin roofing. It was completed in a day and has withstood the use of large Angus beef cattle, sheep, and goats. It has served us well. If you are considering goat shelter options for meat goats, a field shelter may be the type you need. Our cattle went under the shelter when they felt the need, but often stood outside even during snow and rainstorms.

Run-In Shed

Normally, a run-in shed refers to a three-sided open front structure that is facing in the correct direction to protect from the wind. It provides shelter from wind, sun, and inclement weather. Run-in sheds can be simple or elaborate depending on what you decide. The goats will be perfectly happy with a simple, dry windbreak, or a more outfitted building with sleeping platforms, feed troughs, and hay racks. In many cases, the climate will dictate what type of field shelter you will need for your goats.

MATERIALS

- 4 or 5 used pallets in good condition (5 pallets if you want a pallet floor)
- 6 T posts
- Screws and power driver
- Board for climbing to roof
- Shim boards for tread on ramp
- Hammer and nails or drill driver and screws

INSTRUCTIONS

1. If using a pallet for the floor, set in position. Then assemble the three sides

around the base pallet. Alternatively, use only the three pallets for sides and the ground for the floor.

2. Secure the three sides of the run-in shed using pallets and T posts. Using a mallet can help pound the T posts into the ground.
3. Place the roof pallet on the top and secure using screws or nails. If you prefer a more weatherproof roof, you can add a piece of tin to the top pallet or use plywood or boards in place of the pallet.
4. Position the board off one side of the shed. Make the grade as gradual as possible. Secure the board to the roof.
5. Measure the length of the ramp and evenly mark the position of the shims used for treads.
6. Nail the treads into position. Check that all screws and nails are inserted completely into the wood.

Field Shelter

MATERIALS

- 4"x4" sheets of plywood or exterior t-111 siding (4 sheets or 2 full 4"x8" sheets cut in half)
- 2"x4" lumber for framing and roof joists
- 1 sheet of roofing tin
- Nails, hammer, screws, drill driver

INSTRUCTIONS

1. Build the frames for three sides as shown in the photos.
2. Build the front frame as shown.
3. Add the header board to the front frame. This positions the roof at a slight grade for water runoff.
4. Attach the roof joists.
5. Attach the side panels and the rear panel.
6. Attach the roofing tin.

Hoop House Field Shelter

Designed and shared by Ann Accetta-Scott, A Farm Girl in the Making, *and built by Justin Scott. All photos by Ann Accetta-Scott.*

This project will give you a small goat run, or enclosure, which is 4' long x 5' wide x 5½' tall. This blueprint can easily be made longer.

MATERIALS

- 3 (4'x8') concrete reinforcement mesh panels, or 3 (4'x8') cattle panels
- 6 (2"x4") boards, 8' in length
- 3" wood screws, 1 small box

- 1½" wood screws, 1 pound
- 20 3/16 fender washer
- 2 dozen 3" tie wire strips, or 2 dozen medium length zip ties
- Wire cutters
- Bolt cutters
- Impact screw gun with a Phillips head driver
- 1 large tarp, or 1 roll 12'x28' 6 mil Visqueen
- From the 2x4 lumber, cut four 4' pieces, four 3' pieces, two 5' pieces, one 4' 9" piece

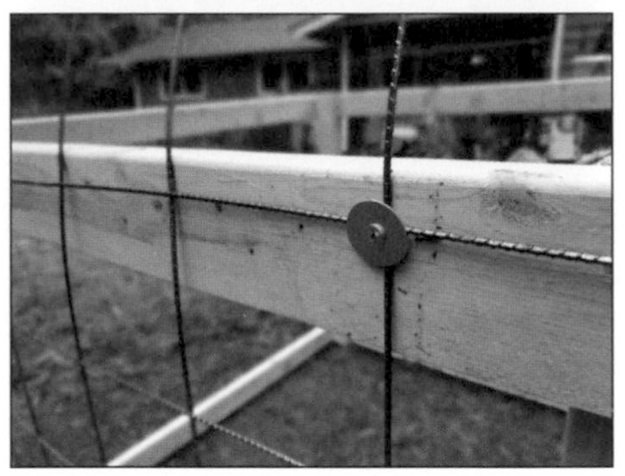

INSTRUCTIONS

The Frame. Build using 2" x 4' lumber and 3" wood screws. Build two side frames: 4' (horizontal) x 3' (vertical)

1. Connect the side frames with the two 5'x2"x4"s to make up the backside
2. Secure the 4' 9" piece to the front end at the bottom

The Top Support (Wire panels, Tie Wire or Zip Ties, Wire Cutters)

1. Using wire cutters, snip 3" tie wire strips.
2. Lay the wire panels end to end to create a 16' piece.
3. Next, overlap the wire panels together by one row, securing the row together using tie wire strips or zip ties every 4".

The Assembling the Run (1½" Wood Screws, 3/16 Fender Washers, Bolt Cutters)

1. With the wooden frame assembled and standing, bend the wire panels over the frame.
2. Secure the wire panel to the frame using the 1½" wood screws and fender washers every 2'.

The Back Panel

1. Stand the third wire panel up on the back side.
2. Secure the wire panel using the 1½" wood screws and fender washers every 2'.
3. Using bolt cutters cut the top to the shape of the arch.
4. Secure the back to the side using tie wire or zip ties.

Applying the Cover

1. Lay the tarp or Visqueen over the fully assembled structure. Keep in mind, the Visqueen can be cut to fit the shape of the frame.
2. To keep the material taunt, fold the corners in and roll any extra material around the ends of the frame. Secure the tarp or Visqueen with wire tie or zip ties every 2'.

Notes

For locations with heavy snowfall, make sure to support the roof. This can be achieved by constructing a ridge support running a 2"x4" front to back, supported diagonally off the vertical side frames.

Barns

Whenever possible, a barn is the best situation for many aspects of goat care. Barns allow the best protection from bad weather. The barn also provides options when caring for an ill or injured goat. Electricity can be added to a barn more easily than to a field shelter. Milking and sanitation are better controlled in a barn, too.

Fencing

The correct type of fencing will keep your goats where they are supposed to be and not roaming your neighbor's flowerbeds. Get your fencing in place before bringing your goats home! We have raised many livestock species and goats are, by far, the most likely to escape and enjoy a holiday. Our first two goats were jumping out of the paddock from day one. That fence was redesigned and rebuilt three times in the first day! The

solution we settled on was post and board fencing covered with either livestock panels or chain link on the outside of the area. This helped prevent the goats from using the wire as climbing steps to freedom.

If you choose to go with a solar electric fence, keep in mind that a few days of no sun may compromise your fence line. Be sure you have a backup plan in place. An alternative to high priced fence boards may lie in used pallets, if you can find them. They can make a nice fenced in pen for some species. You still need to purchase the posts to hold the pallets up.

Goats are quite curious and large gaps in fence boards only serve as an invitation to explore the other side. Split rail fencing is not a good option for goats unless you are planning to attach livestock panels to the inside of the fence line. Remember the saying: "If a goat can put his head through the fence, his body will follow."

Board and post fencing. Goats may try—and succeed at—climbing out without the livestock wire attached.

Electrified mesh fencing.

Livestock grid panels.

Double Wall Fence

MATERIALS

- Fence boards
- Fence posts
- Post hole digger
- String for marking fence line

INSTRUCTIONS

1. Mark the fence line. Measure eight feet from the beginning fence post or building. Install a fence post. Continue installing fence posts.
2. Cut fence boards, if necessary, or order boards cut 8' long.
3. On either side of the fence posts, begin adding fence boards, three in each section, spaced evenly.
4. On both sides of the fence, attach livestock welded wire fencing to the posts. When completed, your fence will be the following layers: wire fencing, fence boards, post, fence boards, and wire fencing. The space between fenced areas should be at least six inches. If the wire fencing doesn't have large open spaces, this should prevent any nose-to-nose touching during quarantine.

Double wall fencing is sturdy enough to contain aggressive bucks during breeding season and is great for when you need to quarantine a goat for any reason. You will use both sides of the fence post, ultimately getting a double fence line using only one set of posts.

What Type of Goat Shelter Do We Use?

Our goat stalls are in the barn, and they open to fenced paddocks. The goats can choose to go out to the paddock when they're not out foraging on the property. We raise a fiber breed called Pygora. These goats grow a fine wool coat that requires shearing twice each year. They are just like other goat breeds in their dislike of any kind of weather other than sunny and dry. The goats will stand at the back door of the stall, leading to the paddock, looking sad and forlorn, if the weather is less than perfect!

When we built the goat barn, we were in the beginning stages of acquiring our goats. Biosecurity is a concern when adding new animals to an existing herd. The interior dividers in the stalls and paddocks were built double to prevent nose-to-nose contact between newcomers and our goats.

Stall Bedding, Buckets, and Feeders

Inside your goat shelter, the bedding should be kept dry and clean. Many goat owners choose to practice a deep bedding method of stall maintenance. This means adding dry bedding to the stall to keep it clean and dry, rather than cleaning out the stall regularly. During the winter, we use this method. It allows a nice deep layer to build up that further insulates the ground that the goats lie on to sleep.

Choosing to clean out the stall weekly or monthly is also an option. I believe it is a matter of personal preference if the ventilation is good, the goats are dry, and there is no odor. Since our stalls are open on one side to the paddock, we rarely have a urine odor in the goat barn. Our challenge is keeping the stalls dry. We frequently check the bedding by picking up a pitchfork full to see if the underneath layers are saturated.

Our water buckets of choice are five-gallon buckets with flat backs. These are placed against the fence line or hung from bucket hooks in the barn. A water trough is a good idea, but the frequent cleaning can lead to a huge water waste. Goats are small compared to cattle and horses, and while all animals require fresh water available at all times, the amount of water a goat actually drinks is much less than a cow's intake. Currently we care for four goats in the goat barn. Two five-gallon buckets is plenty of water each day. In addition to water, some sort of feeding area will be needed for grain, and a way to offer minerals should be provided. More on these two important factors in upcoming sections of the book.

Feeding troughs allow the herd to jockey around for position and, when bumped off the grain by a herd queen, the lower goat can find a new spot to continue eating. Unless you can feed each goat separately (not a normal or easy task), the goats will push each other around to get the grain. Using multiple feed bowls at a time is another solution to goat feeding frenzy.

Goat Sleeping Platform

Building a goat sleeping platform was one of the simplest projects we have completed on the farm. Each section of the goat sleeping platform used two pallets. You can make your goat sleeping platform as large as you need, or to fit what your barn space will allow. I made a double platform for the stall with six goats. They can't all sleep on it, comfortably, but it keeps most of them off the ground. As we reconfigure the barn, arrangements will be made to have sleeping platform space for all the goats.

MATERIALS

For each section of the platform you will need:

- 2 or more pallets in good condition
- 1 sheet of 4"x8" plywood
- Hammer and nails

INSTRUCTIONS

1. Stack two pallets. Add pallet stacks as needed. Two sections of stacked pallets will require one sheet of plywood to cover the open slats.
2. Cover the pallet structure with the sheet of plywood. Use a nail or two in each end to keep it stable.
3. Cover the platform with straw. The space underneath the platform will trap warmer air. Also cover the stall floor with a good layer of dry bedding and straw. Replace wet areas as needed to keep the flooring dry.

Why Build a Raised Goat Sleeping Platform?

Age is one consideration when thinking about building a sleeping platform for goats. Our flock of Pygora fiber goats are getting up in years now. Our first goats, which we purchased in 2004, are considered senior citizens! Older goats may experience stiff, sore joints when they get up from resting. The raised sleeping platform keeps the goats warmer by raising them off the damp ground. This keeps arthritic joints more comfortable, which leads to the goat staying more active. Add a thick cushion of dry straw to the platform to make everything comfortable.

Hoof problems are another reason to build a goat sleeping platform. Anything you can do to keep the goat on dry ground will help to prevent an outbreak of foot scald which, with the right combination of bacteria, can lead to foot rot. Once foot rot is present in your barn or paddocks, it will remain there. It waits for the right opportunity to flare up from a tiny sore area in between the goat hoof "toes."

A third reason to build a goat sleeping platform is because goats like to climb! They will enjoy being up even a few inches off the ground. If the platforms you build are sturdy and stable, the goats will use this structure.

Finally, fiber goats will have a nicer fleece harvest if the goat remains clean and dry throughout the winter. Sleeping off the damp ground helps keep the fiber in top shape.

In Case of Emergency . . .

If your barn happens to get a minor flood from a heavy storm, having a platform already built gives the goats somewhere to stand while they wait for you to "rescue" them. This happened to us one winter. We arrived to find the goats fighting for places that were anywhere above the few inches of water that had invaded the stalls. Building a few swales helped redirect the rain waters, but the goats were very unhappy about the situation!

Manure Management

A small flock of goats, even three or four animals, will create enough manure and bedding waste that you will need to have a plan for how it will be disposed of or composted. On larger properties, without close neighbors, you may be able to have a designated manure pile that naturally degrades and composts the poop and soiled bedding. Add garden compost, coffee grounds, newspapers, and kitchen scraps to keep the pile healthy, and don't forget to turn it regularly with the tractor.

If you have a farm, manure waste can be dried by spreading it out in an open area, and then used as compost on garden beds. Homesteading in a more suburban setting may pose more challenges associated with the manure management. There are removal companies that will pick up full containers and leave an empty container. You may be able to place a three-bin compost system on your property so the manure can mature and age correctly before being added to the garden. While this is the non-glamorous topic of goat keeping, it's good to have a plan before the goats arrive on your property. With suburban farming, being a thoughtful neighbor is part of the process. Keep your goat yard neat and free from fly-attracting manure and your neighbors will be grateful.

Manure Composting System

With backyard farms and smaller properties, a three-compost bin system often does a good job while remaining an attractive part of the yard. With a three-bin system, compostable materials are still added, and the bin contents are turned by moving the contents to the next bin on a regular basis. When the material is in the third bin, it should be well on its way to becoming rich compost for your garden.

Many people allow their chickens to scratch the compost, which aids in turning the material, allowing air to circulate, and the compost process to proceed. If you don't have chickens, using a pitchfork, turn the composting material every week, moving it to the next bin in the line as it starts to decompose.

MATERIALS

- 7 pallets
- 18 screws
- Screwdriver or power drill driver

INSTRUCTIONS

1. For best results, build this structure on or near the area where you plan to use it.
2. Begin by attaching the first side to the first back pallet.
3. Continue to attach the pallets to create a three bin, open front, compost system.

Salt
Let's get Salty

CHAPTER 2: Goat Nutrition

Goat-Proof Food Storage

Yes, I laughed as I wrote this. I don't believe there is a goat proof anything. Goats are smart and inquisitive and rarely give up. If possible, keep stored food out of their reach. Thinking that you will be able to train them to leave something alone is naïve.

Most livestock owners keep feed stored in metal trashcans with tight fitting lids. This is a good start! It keeps rodents from eating the feed and deters the livestock from helping themselves if they get out of the pasture or barn. Taking a step further and building a livestock food locker like the one below will help prevent goats from gorging if they do flip the lid off the metal trashcan.

Food Locker

If you have no other building for storing the feed trash cans, this food storage locker could be placed outside, or against the barn wall. The important part is to never leave the storage locker open. Goats are opportunistic animals and will overeat at a moment's notice. The trash can lids are but one line of defense. The locker is a second, and keeping it locked securely is the third.

The finished box is 58" long by 39" wide and 34" tall. It can hold two 30-gallon metal trash cans for food storage. While not completely rodent resistant, it is well built and does deter rodents from entering and damaging your feed. With the food also enclosed in metal trash cans, this food storage box will keep the goats from getting in the feed and overeating, which is the main purpose.

MATERIALS

Front and back frames:

- 2"x4" lumber for framing the box makes front and back frames (cut 8 pieces 25" long and 4 pieces 48" long)
- 2"x4" lumber for side frames. You will make two side frames from the following cuts of lumber (cut 6 pieces 25" long and 4 pieces 34" long)
- 2 pieces of roofing tin, 56"x36"
- 2 pieces of roofing tin, 36"x34"

For the Lid and Base

- 2"x4" lumber (cut 4 pieces 48" long and cut 8 pieces 25" long)
- 2 pieces of roofing tin, each 58"x39"

Other Materials

- 2 hinges
- 1 metal latch
- Handle for lid

Exterior Finishing Lumber

- 8 pieces lumber, 2'x4'x25"

INSTRUCTIONS

The goal with this project is to put together four metal panels framed in wood, creating a box to hold the trash cans. Another panel will be constructed to act as the lid and will be attached with hinges and a latch for locking. The remaining sheet of metal will be secured to the bottom to act as a weather barrier and rodent deterrent.

This box will hold two galvanized trash cans, one for grain and one for storing the supplements.

1. Build the frames for each side of the box.
2. Build the frame for the lid.
3. Nail or screw the four sides together into a rectangle.
4. Attach the four pieces of roofing tin to the frames.
5. Turn the box upside down and attach the metal to the bottom. Return to upright position.
6. Attach the roofing tin to the lid frame.
7. Measure and mark placement of the hinges for the lid.
8. Attach the hinges to one long side of the lid and then attach the lid and hinges to the box.
9. Attach a latch to the front of the lid.
10. Attach a handle to the lid.
11. Apply the additional 2"x4" pieces as shown to cover the exposed metal edges.

Listeria

Sometimes Listeria is referred to as "circling disease" or "silage sickness." In goats, the primary culprit is listeria monocytogenes. This bacterial pathogen can live in many different mediums, such as moldy hay, feed or silage, feces, or contaminated water or milk, to name a few. Listeria bacteria can also be inhaled from areas with dusty, dry, contaminated feces.

The most common way a goat becomes infected with L. monocytogenes is through eating contaminated rough forage or silage. The bacteria enters through damage to the mucus membranes in the mouth and travels to the brain. Both the L. monocytogenes and the L. ivanovii varieties are zoonotic, meaning they can be transmitted to humans and other animals. Caution and biosecurity methods should be used when cleaning up feces from infected animals, removing placentas and any other body fluids and wet bedding.

The most common version in goats is the encephalitic form from L. monocytogenes. The inflammation of the brain leads to loss of appetite, depressed behavior, fever, and lack of milk. The circling behavior and neck rigidity soon follow. Listeria infection is mainly diagnosed by symptoms or postmortem, due to death occurring in 24 to 48 hours. The mortality is over 70 percent.

Plant toxicity and goat polio can be confused with listeria infection. Goat polio is a result of a deficiency in the thiamine vitamin. Goat polio is not contagious and is treated by injecting B vitamins. Goat polio is more commonly seen in younger kids, and listeria is more common in adult goats.

Treatment for listeria is injections of penicillin, and normally thiamine injections are prescribed also, to cover both possibilities. Anti-inflammatory injections are often given to reduce the brain inflammation. Treatment will require the aid of a veterinarian.

Preventative measures include ensuring plenty of good pasture and barn maintenance and cleanliness. Reduce the access wild birds have to feeding areas, do not feed your goats wet or moldy hay or grass clippings, and reduce stress as much as possible. New animals, new routines, weaning, and parasites are all among stressors that weaken animals.

While no positive natural treatment has been proven, thyme, oregano, rosemary, and garlic are beneficial for immune and healing support. The lichen Usnea (in an extract form) has been shown to target listeria bacteria. These natural treatments have been used by goat owners as a support for (not in place of) the veterinary method of care.

Hay Storage Loft

This is a simple project that will yield enough hay storage for up to 10 bales of hay, possibly more. No one has a good answer as to why goats waste so much hay, but the truth remains, they do. Any hay stored within a goat's reach will be destroyed. Do not store hay in the stall at ground level with your goats. Any hay that is pulled from a stored bale and stepped on by a goat will not be eaten. This is part of a goat's secret code! While it sounds amusing, in reality, all that wasted hay costs you money and time.

MATERIALS

- Measuring tape
- (2) 2"x8"x2' boards for the end supports
- (3) 2"x8" boards cut to the needed length for your stall loft
- Saw
- Screws or nails
- Screwdriver or hammer

INSTRUCTIONS

1. Measure the width of the stall. Use the two 2"x8"x2' boards for either side wall that will support the lengthwise boards.
2. Screw or nail the side boards to the barn walls.
3. Place the lengthwise boards in place.
4. Nail or screw securely.

Tips on Hay Storage

- Store only well-cured, fresh hay that has been dried before baling.
- Store loosely covered in a well-ventilated building or under a weatherproof tarp. Store bales on pallets to ensure good air flow around top and bottom of the stored hay.
- Check stored hay bales before feeding. Look for signs of mold between the flakes, dry rot (old hay), musty odor, and foreign objects that were accidentally baled with the hay. Sometimes pieces of wire, discarded trash, and dead animals can accidentally be baled within the hay.
- In addition to the Hay Storage Loft, using pallets at ground level (away from goat access) and covered by a thick waterproof tarp can be good storage for extra hay.

Legumes and Grasses

Which forage is the best choice for your animals? Goats can have different forage needs depending on what the goat is being kept for.

Hay comes from two main growing sources, legumes and grasses. Straw can be used as a forage if cut young and before the grain is harvested. Most people buy bales of straw with the intention of using it for bedding the stall. We use straw as bedding, although our goats have been known to munch on a straw snack.

Legumes include alfalfa and clover, two popular choices for energy and nutrition. Alfalfa hay can contain almost two times the protein and three times the calcium of grass hay. Usually the protein range for alfalfa is 16 to 18 percent . Protein content will vary depending on when the hay is cut and the maturity of it at the time of cutting. Some animals, such as a dairy goat or a dairy cow providing milk for your family, may need the higher calorie and higher protein content provided by alfalfa hay. But for an animal that just needs forage to keep the rumen functioning and to provide some calories and nutrition, a grass forage may be the better choice.

Grass hays include bromes, fescues, orchard grass, rye grass, wheat grass, timothy, coastal Bermuda or bluegrass. Hay from grasses contains fewer calories (energy) and a lower protein amount, usually 6 to 10 percent. Your feed store may carry a mixture of alfalfa and orchard grass, or a mixture of timothy and orchard grass. If your animals need a certain level of protein or energy from their hay, it is a good idea to ask what the bale consists of. I do not recommend feeding alfalfa unless the animal requires the higher energy and nutrition.

Forage from Grains

Grains, such as oat straw or barley straw, can be fed as a nutritional feed if it is cut young and still has the grain intact. After the grain is harvested and the stems are mature and woody, it is called straw, and the main purpose will be for bedding use or ground cover. Straw is usually less costly than hay and has minimal nutritional value.

Forage for Animals with Special Nutritional Needs

Lactating animals, including nursing mothers, are often fed an alfalfa hay to keep them from losing condition during their milk production. Whenever you need to switch from a grass hay to a legume hay, do it gradually. The type of feed given contributes to the acid balance in the digestive tract. Switching abruptly can cause bloat or acidosis in your animal.

Young kids that are growing rapidly and have tender mouths may require the softer, leafier alfalfa. If they cannot chew the tougher stems, their nutritional needs will not be met, and growth and development will be challenged.

The best way to check the hay you are going to feed is to cut open the bale and inspect it. Moldy hay has a distinctive odor. Look at the hay. It should be fresh looking and include leafy material along with stems. Weeds are not a problem, and goats especially will enjoy them, but make sure that the weeds are not from any toxic plants.

Hay is generally packed in round or square bales. Hay should be stored out of the weather in a well-ventilated, covered area. We store our hay in an open front shed. Every few days we will bring over a few bales and stack them in the barn for convenience. We use an orchard grass hay for our goats as we are not breeding and have no lactating animals currently.

The bottom line on feeding hay is to know the content of the hay, look at the condition of the hay, and use the appropriate hay for your animals' nutritional needs.

Hay Needs

Estimating how much hay you will need is tricky when you first acquire goats. Different stages of life, pregnancy, and lactation are going to play into the amount of forage required per day. If your goats are only being fed hay as their forage requirement, estimate upwards of 5 percent of their body weight per day, with lactating does requiring more.

Goats are wasters of hay. As soon as a bit of hay, pulled from a rack, hits the ground, it is useless in their eyes. Not to mention the hay that the next goat has accidentally peed on or walked through. Keeping the hay off the ground as much as possible helps conserve hay and waste less, but the truth is, goats are big hay wasters.

My older Pygora goats, a medium-sized fiber breed that are being kept strictly for fiber these days, eat approximately two flakes of a square bale each, during the course of the day. When I say "eat," I should say I provide that much for each goat. Much of it seems to end up on the barn floor, but they don't look hungry, so that is my best guess.

Judging Hay Quality

The quality of the hay is a factor in how much you need to feed the goats. While any goat keeper will tell you that goats waste hay, they still need enough nutrition from the forage. If you restrict hay because they waste it, you might be restricting their nutrient levels.

What to do?

Provide plenty of free choice hay or forage. When feeding hay, use a small opening hay feeder. When using a traditional livestock hay manger, add wire to the inside of the manger. This will restrict the goats from pulling out large chunks of hay, only to let it fall to the barn floor. Once they stand on the hay, there is little chance they will eat it.

Buy clean, dust-free, debris-free hay from a reputable source. Some hay is only a mixture of dried orchard grass or weeds. While the goats may be ok with it, if the hay has been stored for a long time, it may lack the proper nutrition.

The following examples and projects are simple methods used to allow free choice forage for your goats, while preventing too much wasted hay.

Stair Railing Hay Rack

This is one of the simpler ways to build a sturdy hay rack for your goats. There is one piece required! Scrap pieces of 2”x4” lumber can be used to brace the railing at the angle required. Fasten with screws for the best durability.

MATERIALS

- 1 new or used stairway railing
- Screws and drill driver to attach the railing
- Scrap pieces of 2"x4" lumber for support braces

INSTRUCTIONS

1. While holding the railing, mark the approximate location where it will be installed.
2. Using screws or brackets and screws, attach the lower edge to the wall. It will be helpful to have a second person hold the railing during this installation.
3. Allow the railing to fall away from the wall slightly on the upper edge. This will allow hay to fall between the wall and the railing, acting as a hay rack.
4. Measure the distance needed for the upper braces and cut the 2”x4” lumber to correct length.
5. Attach the 2”x4” support brace to the barn wall, and then to the upper edge of the railing, at both ends.

Pallet Hay Rack

This is a good recycled pallet project for offering your goats hay. The small space between the pallet board helps prevent too much waste. The pallet will accommodate two goats at a time. Hang at the right height for your herd.

MATERIALS

- ½ of a full-size pallet
- Scrap lumber for sides of the hay feeder
- 3 boards, each 12" long (scrap lumber is fine)
- Hammer and nails or screws and drill driver

INSTRUCTIONS

1. Hold the pallet up and mark the height for installing.
2. Attach the scrap lumber pieces first to the pallet, and then attach the lumber to the barn wall at an angle.

Photos by Ann Accetta-Scott

Free-Standing Hay Rack

Project and photos by Mandi Chamberlain

MATERIALS

- 9 pieces of 2"x4"x4' lumber
- 12 pieces of 1"x2"x 3' lumber
- 4 pieces of 1"x2" lumber cut as follows: 2 pieces 18" long and 2 pieces 12" long.
- Drill driver and 2" wood screws

INSTRUCTIONS

1. Begin by making the three X braces.
2. On one side at a time, attach the top rail. Stand the hay rack up.
3. Attach 6 rails evenly spaced on each side.
4. Attach the end braces.

Plastic Storage Bin Hay Rack

This is an inexpensive project that works well for a small herd of two to four goats. It might be fine for larger herds too, or you could install more than one of these. It helps keep the hay clean, and the goats can't pull out too much hay at one time, then wasting it once it hits the ground.

MATERIALS

- 1 large plastic storage tote (you won't need the lid, so if you have a tote hanging around just use that)
- X-Acto knife to cut out the eating holes
- Scrap boards for attaching the bin to the barn wall or fence
- Drill driver and screws
- One piece of ¾" plywood big enough to cover the tote
- 2 metal hinges

INSTRUCTIONS

1. Measure and place the four eating spaces evenly on the front and two sides of the bin. I made four holes that are 4" x4" and spaced them on the front and one side of the tote. Mark these areas with a sharpie marker and then cut out the spaces.
2. Cut three pieces of scrap lumber that will fit into the tote. These will be fastened to the barn wall or fence board to stabilize the tote.
3. The plywood lid will be attached separately, for both safety and air circulation. If you install this hay feeder outside, be aware that this lid will not keep rain from getting into the hay in the tote. Keep dry hay available and clean out any wet hay promptly.

DEWALT

Pasture or Paddock

Rotational Grazing

The amount of pasture you have available for grazing is finite. Stocking rates may vary based on the quality of the forage being grown. It's best to keep your herd numbers lower than the maximum in case of drought or other natural circumstances that would limit your forage growing.

Rotating pasture allows the grass to grow taller between grazings and keeps the goats from eating the grass close to the ground where the parasites are lurking. In addition, rotating pasture breaks the parasite life cycle by removing one of the intermediate hosts, your goats. Without a host available, the parasites lie in the pasture and die.

Rotating pastures has less impact on the ground. Daily travel over the same pasture will eventually compress the soil and make it harder for regrowth. It also allows the manure to be spread more evenly over the property to fertilize a larger area.

Rotating and allowing regrowth over many separate paddocks increases strong pasture growth and allows for a longer grazing season.

Finally, rotating animals through the different paddocks gives you a great chance to assess their condition and behavior and creates a friendlier animal.

In short, here are three main reasons to consider rotational grazing:

1. Improve pasture quality
2. Reduce feed and hay bill
3. Maintain a healthier herd

Controlling Toxic Plants

Your goat can exhibit symptoms that range from mild to life threatening after eating a toxic or poisonous plant. Milder symptoms could include some mouth frothing, excess salivating, weakness, diarrhea, and depression or self-removal from the herd. Depending on the degree of poisoning, the symptoms may progress to serious life-threatening issues such as convulsions, rumen issues, paralysis or other motor coordination problems, respiratory distress, and circulation complications. With small backyard goat herds, the concern is with common ornamental plants and landscape shrubs. An extensive list of toxic plants can be obtained from an Internet search or from your local agriculture extension office.

Some common plants that can cause toxicity issues in goats include:

- Avocado
- Azalea
- Belladonna
- Blue Cohosh
- Boxwood
- Buckeye
- Buckwheat
- Burning Bush Berries
- Buttercups
- Castor Beans
- Cherry
- Choke Cherry
- Clover
- Crowfoot
- Elderberry
- False Tansy
- Hemp
- Holly
- Horse Nettle
- Japanese Yew
- Johnson Grass
- Laurel
- Lilac
- Lily of the Valley
- Lupine
- Marijuana
- Milkweed
- Mountain Laurel
- Oleander
- Pine Trees (in great quantity, see section on "Can Goats Eat Christmas Trees," page 50)
- Poison Hemlock
- Poke Weed
- Ponderosa Pine
- Red Maple
- Rhododendron
- Water Hemlock
- Wild Black Cherry
- Yew

*List is not all inclusive

If your goat is suffering symptoms of poisoning, contact your veterinarian immediately.

If you have activated charcoal on hand in your emergency first aid box, administer by mixing with warm water and orally drenching the goat. There should be amounts listed on the container based on the animal's weight.

DIY Weed Killer

A healthy pasture of good forage grasses will choke out weeds and make the ground less hospitable to toxic weeds. In the meantime, how do you control the growth of plants that might make your goats sick? Spraying chemical weed control products adds toxic chemicals to the very plants that you want your goats to eat. Here are two recipes for non-toxic herbicides that can be safely used around livestock. These are intended to be directly applied to the plant you want eliminated. These sprays are not intended as a broadcast spray as they will kill any plants they are sprayed on. Salt, Epsom salt, boiling water, and straight vinegar will also act alone when applied directly on an unwanted weed. In addition, mechanical means of smothering toxic weeds may be a solution. Consider smothering with mulch or newspaper or digging up the plant in question and disposing of it.

Weed Killer Recipe 1

INGREDIENTS

- 1¾ cups white vinegar
- ¼ cup castile liquid soap
- 3 drops of each of the following essential oils: wintergreen, clove, cinnamon, and orange

Weed Killer Recipe 2

INGREDIENTS

- 1 gallon white vinegar
- 1 cup Epsom salt
- ¼ cup castile liquid soap

INSTRUCTIONS

1. Combine all ingredients in a spray bottle. Hint: use a funnel for easier filling.
2. Shake to mix all ingredients completely. Shake before each use.
3. Spray directly on the weeds you want to kill.

Pasture management includes walking the pasture to note plants that can cause health problems for your herd. If you have a very large property this can take considerable time, but the management practice will pay off in better health for your herd. In addition, it is my belief that when goats have plenty of choice in forage available, they will not willingly eat toxic plants. Keeping your paddocks healthy and thriving should limit the threat of an occasional toxic plant.

Grain Needs

Grain is not always a necessity for goats. A field of wethers kept solely for weed control can forage for the nutrition they require, or you can supplement with a good clean grass hay. Maintenance of a non-lactating, non-breeding flock does not have to include a bag of goat grain. That said, there are some advantages to keeping the goats familiar with the treat of sweet grain. When your goats realize that the feed bucket means a yummy treat is ready, they are more likely to follow you back to the barn.

There are good reasons for having a balanced ration on hand for your goats. Growth, pregnancy, and lactation are three times when your goats will benefit from a limited grain feeding. Training is another good time to have grain on hand. We have always trained our livestock to the sound of the feed bucket. It brings them running from out of sight distances. They love a treat of goat grain, and it's a good way to lure them back after a fence breach. Grain is useful for conditioning animals that might not thrive on your local browse or hay. For animals intended for breeding or showing, a balanced grain ration will fill out the goat and put on some weight. Our goats get some grain while they have their hooves trimmed, and while being sheared. Grain makes everything sweeter.

Because of it's nutrient and caloric density, grain should be measured out and securely stored where the goats can't have access to it. Overeating grain can cause bloat and death.

Formulating a Grain Ration for Goats

Goats do best on a forage-based diet, including browse and hay. Grain rations or concentrates should be offered for increased calories, lactation, late gestation, and growth. When formulating rations for any animal species, balancing the ingredients is the most important factor. Premixed, commercial rations are prepared under the guidance of knowledgeable animal husbandry experts. Much research has gone into these diets, although they do sometimes contain individual ingredients that we prefer not to use with our animals. Whether you have an allergy to certain grains or just have taste preferences, what you feed your goats will to some degree pass through to the milk and meat. It is possible to create a balanced ration using individual grain and supplement ingredients.

Check with your local grain mill or feed store for the availability of certain grains before beginning. Select a rodent-proof storage container with a tight-fitting lid. Before committing to feeding a do-it-yourself ration, be sure that you have the time to do so and can regularly obtain the ingredients. Suddenly switching between goat feed mixtures is not a good idea.

Before offering my suggestion for a do-it-yourself goat concentrate, please note that there are as many ideas for making a goat ration as there are goat owners. Every region of the world will have different ingredients that are more readily available. Consult your local grain retailer or grain mill for availability before making a final decision. In addition, the decision to feed corn or soybeans as a nutrient source is a personal decision. If you choose to not use these foods, look for an alternative with a similar nutrient composition.

Grain Mixes

I am presenting two different mixtures, one that uses corn and one that does not. A Google search will yield many other possible mixtures. Educate yourself about what nutrients are dense in your region and what nutrients are sparse or lacking. A one size fits all ration for goats will only meet the baseline of a goat's needs. You can feed for your specific herd, their purpose, and their season of life with a bit of research and education.

Feed Ration #1

INGREDIENTS

- 1 part split pea or lentils
- 1 part flax seed or black oil sunflower seeds (BOSS)
- 1 part barley
- 1 part oats

INSTRUCTIONS

1. In addition to the grain mixture, offer baking soda, salt, and minerals, or a premixed commercial mineral mix. My suggestions for minerals are on page 45.

Feed Ration #2

- 1 part oats
- 1 part alfalfa pellets
- ½ part corn
- 2 tablespoons brewer's yeast
- 2 teaspoons kelp
- ½ cup black oil sunflower seeds (BOSS)

INSTRUCTIONS

1. Add molasses if needed. Molasses adds calories, and palatability. Some goats love it, and others would rather you left it out. Caution, though, with adding molasses: it also adds moisture. During the hot weather seasons, moisture might lead to mold, so keep an eye on the feed and do not mix up too much at one time.
2. Include free choice baking soda, salt, and minerals as discussed above. See page 46 for more information and a Do-It-Yourself Mineral Mix formula.

Salt, commercial mineral blend and ground kelp, and baking soda.

How Much Grain/Feed Does My Goat Need?

Here's another question that can only be answered with a general guideline. In addition to making sure your goats have enough energy, you will want your goat to be fed enough to look healthy without becoming overweight. Evaluating condition in your animals is a skill you will develop over time. Start by learning to feel along the top line of your goat. A healthy weight goat will have a slight round to the top line. Each breed has certain standards for appearance. Learning these will help you determine if your goat is getting too much concentrated feed.

The following guidelines are very general:

- Milking does—1 cup of grain 2 times a day during milking
- Growing kids—¼ to ½ cup of grain per day
- Wethers, and non-bred does—small amounts as a treat. It is often helpful to have goats trained to the feed bucket. This helps to get the goats to follow you if they think a tasty treat is at the end of the walk. A little grain sprinkled on the ground once you get them in the new area is a sweet reward.

Feeding time can be a bit chaotic when you are feeding a small herd of goats. Grain helps the body stay warm during cold nights and keeps weight on older animals that don't forage as well as they used to. In addition, your breeding animals will benefit from grain feedings, and lactating does absolutely need the extra calories.

Recycled Feed Trough

For this project, you can reuse an unwanted sliding board, particularly the sliding boards made of heavy flexible plastic. The plastic slides last for years, and many families have no use for them once the children have grown. Check local yard sales, reclaimed stores, Goodwill, and on-line marketplaces for these finds. Your measurements may vary based on the size of the sliding board. I am providing the lumber sizes we used.

MATERIALS

- Used sliding board
- Support cross legs
- 4 pieces of 34"x2"x4" lumber
- Base
- 2 pieces of 74"x2"x6" lumber

Goats eating out of sliding board.

- 2 pieces of 28”x2”x6” lumber
- Side support for slide
- 2 pieces of 87”x 2”x4” or 87”x2”x6” lumber
- Hammer and nails or drill driver and screws
- Tape measure
- Pencil for marking
- Hand saw or circular saw

*Note: Measurements will vary based on the sliding board being used for this project.

INSTRUCTIONS

1. To build the legs: using only one screw or nail at first, assemble the cross sections for legs.

2. Place the sliding board on the legs to check for the angle needed. Adjust the angle for steady support. Once you are happy with the placement, insert additional screw in the leg assembly to secure the position.
3. Build the base by attaching the 74”x2”x6” boards to each 28”x2”x6” board to form a rectangle.
4. Attach the cross-leg assemblies to the base. Attach the side support boards.
5. Position the sliding board. We did not attach the sliding board. This enables us to remove it for cleaning or moving to a new location.
6. If your sliding board feeder is located where it can collect rain water, drill small holes through it for drainage.

Wooden Manger Feeder

With just a few boards, you can put together a wooden feeder for your herd. Use the following guidelines and build a feeder of any size to fit where you need it. The feeder trough described here comfortably accommodates six or seven goats.

SUPPLIES

For the Sides:

- 2 pieces of 74"x 2"x6" lumber

For the Base:

- 1 piece of 74"x2"x12" lumber

For the Legs:

- 2 pieces of 18"x1"x10" lumber (Note: the "legs" are also the ends of the manger)
- Handles (ours were made from scraps and attached for ease of carrying the feeder)
- Drill driver, wood screws, drill bits for pilot holes

INSTRUCTIONS

1. Assemble the sides and the base. At each end, attach the leg boards. Attach scrap lumber for carrying, if desired.

Tire Feed Bowl

Feeding goats, especially a small herd, can get interesting. Herd hierarchy is a factor, along with the fact that goats are greedy come feeding time. A tire feeder can eliminate some of the drama by keeping the pushing to a minimum due to the round size. The goats will still push each other around the tire though!

A tire also prevents the goats from picking up the feed bowl and dumping the food on the ground. Ground feeding can increase the risk of parasites.

MATERIALS

- 1 or more used pickup truck–size tires
- Large rubber feed bowl that snugly fits inside the tire

INSTRUCTIONS

1. Push the feed bowl snugly into the tire opening. That's it! Not all projects have to be difficult in order to solve a problem!

Minerals, Salt, Baking Soda, and Supplements

These supplements are often fed free choice from small feeders attached to the stall wall. Feeding minerals helps balance out the goat's diet since they often are limited to what browse they can access.

Just as with commercial grain rations, there are many formulas available commercially to add minerals to your goat's diet. You also have to decide whether to mix the minerals into the grain or offer them free choice. Surprisingly, not all the mineral mixtures contain the same percentages of ingredients. I looked at a variety of goat mineral supplements and found varying percentages of calcium, phosphorus, salt, selenium, copper, zinc, and magnesium. Educate yourself about the minerals available in the soil in your region. Find out which deficiencies are common.

Our goats receive their mineral supplements using a free choice mineral feeder. We have a separate feeder for aluminum chloride (to help prevent urinary calculi), and yet another for baking soda (for healthy rumen activity). During the hot summer months, free choice livestock salt is also available.

For the mineral mixture, I have chosen kelp as the main source. Kelp, or marine brown algae, is also an immune system support, aids in relieving stress, and increases meat quality and weight gain. It's rich in many vitamins necessary for good health.

Here's a list of some of the nutrients provided by kelp:

- Calcium
- Iron
- Magnesium
- Phosphorus
- Potassium
- Sodium
- Zinc
- Copper
- Manganese
- Selenium

Do-It-Yourself Mineral Feeder

PVC pipes are sturdy, non-toxic, and easy to clean. They make a good choice when building DIY projects for livestock and farm needs. The mineral feeder shown here is easy to set up and refill. The feeders hold enough mineral and supplements for a small herd of goats, but not so much that you must worry about stale product.

MATERIALS

- 4" 90-degree U-shaped PVC pipe
- Drill driver and 1" wood screws
- 4" test cap or a 4" clean out plug
- Scrap lumber for base

INSTRUCTIONS

1. Position the pipe in a way that keeps the minerals or supplement from readily pouring out of the tube. Mark the position.
2. Using a drill driver, a drill bit, and 1" screws, hold the tube in place while drilling a pilot hole for the screw. Insert the screw in the lower section and repeat the process with the upper section. See photos for location of screws.
3. Repeat the procedure for the other two pipe sections.
4. Hold the assembly at a good level for all your goats to easily access all three pipes. Screw the base board to the barn wall, top and bottom.
5. Fill the pipes with your choice of supplements for your herd. I used brewers' yeast, baking soda, and livestock salt. Ammonium chloride, sheep mineral blend, or an herbal blend are other possibilities for these feeders.

Wooden Bucket Mineral Buffet

This DIY project was inspired by a trip to the local big box home improvement store. I found a wooden shelf and the three buckets, which I think were intended as planters. Instead I saw a goat mineral buffet. For the more creative goat farmers, you could paint this in many artistic styles and add a touch of whimsy to your goats' daily life. I kept the decorating simple, but the idea is there.

MATERIALS

- 1 premade unfinished shelf
- 3 small buckets or pots
- Scrap lumber for attaching the feeder to the barn wall

INSTRUCTIONS

1. Paint the shelf (optional). I did not paint the buckets that will hold the supplements because of the potential for goats nibbling on the buckets.
2. Place the shelf against the barn wall and mark the placement with a pencil or marker.
3. Using wood screws, attach the shelf to the barn wall.
4. Insert the bucket and fill with minerals, baking soda, and livestock salt.

Paint Bucket Mineral Feeder

Here's another project inspired by walking through the mega home improvement store. I found three empty paint cans. The lids came with them, but I don't need the lids for this project. I picked up the three wrought iron hooks at a flea market.

MATERIALS

- 3 small paint buckets or other small buckets with adjustable handles
- Hooks to hang the buckets from the stall or barn wall

INSTRUCTIONS

1. As with the mineral buckets in the previous project, decorate or paint the paint cans if you wish.
2. Measure the spaces to attach the hooks to the barn wall.
3. Install the hooks on the wall.
4. Hang the buckets and fill with the needed supplements

Block Mineral Holder

Project from Elizabeth Taffet, Garden State Goats

This mineral block holder is installed near the ground. Place it out of the elements, such as in the stall or under a roof overhang. It is a combination of a homemade box that will hold the mineral block and a two-compartment purchased loose mineral holder.

MATERIALS

- Scrap pieces from 1"x2" lumber
- 1 square piece of plywood, large enough for the salt block to rest upon
- Nails
- Hammer

INSTRUCTIONS

1. Measure the mineral block you commonly purchase.
2. Using 1"x2" lumber and the plywood, build a small box that the mineral block can sit in.
3. Using additional scrap lumber, attach the box to a post in the building and use a scrap of lumber for support.
4. Attach the loose mineral feeder to the front of the assembly.

Can Goats Eat Christmas Trees?

Purchasing a fresh cut tree from a local tree lot or cutting your own Christmas tree is a fun holiday activity. After the tinsel and ornaments have been removed, using the tree as a food option in the barnyard can add extra value to a fresh cut tree. As with any supplemental snack, moderation is the key. If you have several fresh Christmas trees available, hold some back for another day.

The genus Pine contains a lot of plants, some not even true pines. Yew is not in the genus of Pinus (it's a member of the Taxus genus). Yew is often confused with pine but can cause toxicity and illness in most animals. Many of the popular varieties selected as Christmas trees can be used as a food supplement in limited quantities. White pine and Scotch pine are common, along with the Fraser fir, Douglas fir and blue spruce. With any edible, I never recommend overfeeding. Illness can result just from the upset in the diet routine. Stick with the adage of "everything in moderation."

Pine needles provide trace nutrients, antioxidants, minerals, and forage. Trees should not replace the normal forage, grain, or other feed material. Pine is good for intestinal worm control and adds Vitamin C content to the diet. Some varieties contain higher amounts of Vitamin A, too. In addition, the activity of chowing down on a tasty novelty interrupts the boring days of winter and the monotony of eating only hay.

Pine needles can cause abortion in cattle, if eaten in varying quantities. Although cattle and sheep and goats are all ruminants, the absorption mechanisms in cattle seems to have more of a problem with pine. Problems have been documented with Ponderosa pine, Lodgepole pine, and Monterey pine showing a link to premature birth and abortions in cattle. The yew is another member of this group that can be extremely toxic. Horses and ponies can colic from too much pine. I include this in case you are running different species together in the same area. Chickens will also enjoy munching on some branches of the tree, although hopefully you are not keeping your chickens and goats in the same paddock.

If toxicity is a potential issue, why consider feeding the pine tree to the goats? For starters, one tree per small flock of ten to twelve animals isn't enough to cause toxicity issues. Goats eating nothing but pine bark, branch tips, and needles every day can lead to toxicity and abortion, along with other health risks. If other family members, friends, and neighbors drop off their used Christmas trees for your goats, limit them to one tree at a time. Although it's better to be safe than sorry, I have found that with most toxic plants, goats will get full before they eat so much that they experience toxicity issues. If the toxic plant is the only choice, the ruminant will eat it, and that can cause problems quickly. If there are plenty of other nutritious foods available, the animal will not normally choose to eat the toxic plant. In short, a small amount of pine Christmas tree will add nutrients and not cause harm to your herd.

What about the toxicity of man-made products applied to Christmas trees? This is a topic that always comes up when talking about feeding fresh cut trees to livestock and poultry. Some large retailers still apply a fire-retardant spray or a colorant to the trees. Ask your seller about this. Often, the fire-retardant spray is a green color that can be seen on the trunk and some of the branches. The tree may appear to be a brighter green than you would expect. In any case, ask questions and inspect the tree carefully if you are planning to feed the tree to your farm animals.

Check with your seller before assuming that a tree is all natural. If you buy your tree from a small independent lot, they should know where the trees came from and how they were prepared for sale. If you cannot be certain, don't feed the tree to the goats.

Goats can keep your Christmas tree from ending up in the landfill. There are healthy nutrients in the tree, and feeding it to your barnyard animals is safe in occasional small doses.

Photo by Elizabeth Taffet, Garden State Goats.

Treats for Goats

Molasses and Oats Goat Cookies

Makes approximately 4 dozen small cookies.

INGREDIENTS

- 1 cup granulated sugar
- ½ cup butter
- 1 egg
- ½ cup molasses
- 1 teaspoon vanilla extract
- 2 cups flour
- 1 teaspoon salt
- 1½ teaspoons baking soda
- 2 cups oatmeal

INSTRUCTIONS

1. Preheat oven to 375°F. Grease a baking sheet or use parchment paper to line the baking sheet.
2. In a large mixing bowl, cream together sugar and butter.
3. Add the egg and molasses and mix well. Add vanilla extract and mix.
4. Combine the dry ingredients except for the oatmeal. Add the dry ingredients into the creamed sugar mixture. Stir to combine.
5. Add the oatmeal and mix well.
6. Drop teaspoon-sized amounts of dough onto the baking sheet. You can roll them into balls or flatten into cookies if preferred.
7. Bake for 10 to 12 minutes, or until baked through.
8. Store in an airtight container.

Bar Cookie Variation

The Molasses and Oats recipe can also be baked as a pan-style cookie. Grease or line a 9 x 13 baking pan or two smaller pans. Spread the cookie batter in the pan, smoothing and making it as even as possible. Pre-score the cut lines for the cookies. I made the cookies approximately 1-inch square.

Bake the cookies at 350°F for 25 to 30 minutes. Some ovens may require a longer bake period. Test the center for doneness.

Cool the pans completely before finishing. When cool, separate and complete the cutting. Store the softer cookies in the freezer for future use.

Herb and Oatmeal Treats for Goats

This treat is a round ball, rolled in oatmeal and then baked. Since they are a softer treat, store them in the freezer for longer-term storage. Though the goats might convince you that there's really no need for longer term storage!

INGREDIENTS

- 2 teaspoons cinnamon
- 2 teaspoons turmeric
- ¼ cup fennel seed
- ½ cup dried mixed herbs (I used basil, oregano, parsley, thyme, sage, and rosemary)
- ½ cup solid coconut oil
- 1 egg
- 1 teaspoon vanilla extract
- ½ cup molasses
- 1 cup all-purpose flour
- 1 cup oatmeal

INSTRUCTIONS

1. Preheat the oven to 350°F. Line a baking sheet with parchment paper or lightly grease with coconut oil.
2. Mix the herbs, cinnamon, turmeric, and fennel seed together in a medium sized mixing bowl. Add the coconut oil, egg, vanilla extract, and mix.
3. Add the molasses and flour, and mix thoroughly.
4. Place the oatmeal in a shallow bowl.

5. Taking 1 teaspoon of dough in your hand, roll into a ball. Roll ball into the oatmeal to coat. Place on the cookie sheet. Repeat until all dough is used.
6. Bake for 20–25 minutes (check after 20 minutes since ovens can differ).
7. Store the treats in an airtight container in the freezer if you won't use them immediately.

Can Chickens and Goats Live Together?

Many small property owners ask this question. When space is limited, how do you know which animals can safely be kept in the same paddock or barn? Goats are friendly animals and our herd members are usually more curious than cautious about other barnyard species. Although the goats do not mind an occasional meeting with the barnyard chickens, we do house them and feed them separately.

Feeding is a major concern with goats, and they will happily overeat all the chicken feed, if given the chance. This can lead to life threatening bloat and rumen problems, but also, the chickens will quickly go hungry! Our chicken feed is kept in the coop or fenced run, away from hungry goats.

Disease transmission is another factor when allowing chickens and goats to co-mingle. Chickens are happy to leave droppings everywhere they travel, even in the feed bowls and water containers. A few parasites can be transmitted between the species leading to illness in your goats.

Cryptosporidium has a few different types, and some can be transmitted to small ruminants and also humans. When chicken droppings contaminate a water bucket, the parasite can be transmitted to the goat. Eating food from a bucket, or off the ground near chicken droppings, can also transmit cryptosporidium. When this microorganism gets out of control it can sicken your goats and your chickens, causing intestinal bleeding, severe diarrhea, and respiratory complications in poultry.

Other zoonotic diseases can be transmitted from chicken droppings to your goat herd. Campylobacter and salmonella are opportunistic bacteria that can cause illness even in low levels.

These are just a few of the possible contaminants that can cause herd health problems if chickens and goats are housed together. Occasional wandering of the barnyard at the same time shouldn't pose a significant risk, but housing the species separately is important for avoiding the spread of disease.

Dr Naylor
Veterinary
Topical Antibacterial
HOOF 'n HEEL
molasses
oatmeal
goat cookies

CHAPTER 3: Goat Health

Coccidiosis

Coccidia are small protozoa. In a normal, healthy herd, a small amount of coccidia will not make your animals sick. But as is the case with many opportunistic infections, the parasite looks for an opportunity to strike, and in the right conditions and a weakened physical condition, illness will occur. This is a common, and often self-limiting illness. If caught early, the animal is treated and recovers quickly.

Young ruminants on pasture are more susceptible to cocci than adults. Adult animals will most likely have some oocysts, but they often have developed immunity. The normal activities that often are part of the life of young ruminants can contribute to them having an episode of coccidiosis. Shipping the animals can lead to lack of feed being ingested, which can lead to a weakened animal. Also, chilling, stress, crowded conditions, and new surroundings will make a young animal weakened and more susceptible.

Crowded conditions play a big part in coccidia infections. Over-grazing pastures and failure to rotate grazing areas are other contributing factors. The protozoan must live for a part of its life cycle outside of the host animal. Grazing on very short or sparse grass will cause the animal to eat more of the oocysts that have been shed in the feces of the flock. Animals that are susceptible due to the conditions mentioned above will quickly take on too many oocysts that will mature inside the muscle tissue, stomach, and small intestine, and cause illness. If you are keeping goats in a small stall and paddock, cleanliness if even more important.

Goats infected with coccidia will have symptoms such as failure to thrive, mild diarrhea, loss of appetite, and anemia. The condition can worsen until the animal nears death from the symptoms.

Cocci need time for the oocysts to live outside of a host. Sunlight and dry conditions prohibit the eggs from developing and may keep the outbreak contained. During wet, rainy, and humid weather, the parasites can live a long time.

In an animal with an acute infection, antibiotics are prescribed. Coccidiostats may help prevent an infection but not necessarily take care of an existing infection.

Keep in mind that adult animals can become infected, too, it is just less likely. Circumstances such as a stressful delivery, another illness, or unsanitary conditions can cause any animal to have an overload of cocci.

The research is not conclusive as to whether other animals can become infected from a goat with coccidiosis. It used to be thought that the cocci were species specific. Some research has shown that eating an intermediate host such as a rodent can lead to infection. When there is any risk or question, taking a fecal sample to a veterinary lab for analysis is a good idea.

Natural Wormers and Natural Treatments

I have never subscribed to the practice of treating for things unless I have reason to believe an animal is sick or is showing symptoms. I prefer to take a natural, preventative approach to parasite overloads.

Parasitic worms are present in the environment. Once we introduce livestock into the picture, we add an available host for the life cycle of the parasite. If pastures are not rotated, or if the animal load is too high for the property, the parasites have a strong breeding ground for infesting your goats. Manure left on the pastures in too high a concentration begins to be a breeding ground for the larvae stage. As the larvae develop, they migrate to the plants, where they are unwittingly munched up with the foraged greens. The goat becomes the unwilling host while the larvae develops, allowing it to feed off food being digested, or blood. Most animals that graze or forage will have some intestinal parasites. The illness from internal parasites occurs when the load becomes detrimental to the animal's health and interferes with it getting enough nutrition. This is more likely to happen if an animal is stressed, and therefore more vulnerable to health issues of all kinds.

Livestock can exhibit diarrhea, anemia, rough unkempt coats, weight loss, and sudden death. Worms can cause dehydration and weakness and a general failure to thrive. Undiagnosed parasite illness can certainly lead to death.

Making the decision on whether to use a natural remedy for your farm animals is a personal choice. Although I prefer to stay with natural cures, there are times when it's in the best interest of your animal to use a medication. If an animal is already sickly from a parasite infestation, you may not be able to save it with natural means. You need to decide based on what you observe. Often, when a person brings home new animals, they don't realize that worms are the cause of the animal not thriving. When the diagnosis is made, it may be too late. I take an all-inclusive approach to parasite management. If I notice an animal is not doing well on our regular course of natural treatments, it is a high priority to get a fecal sample to my vet for a more conclusive diagnosis. While I do believe in the power of natural preventatives, they work better when used as a preventative, along with good pasture rotation management. When the goats are already ill from parasites, I choose to use a chemical wormer to bring them back to a healthy status as soon as possible. It is equally important to learn which worming medication is appropriate for the different types of parasites. Sometimes you can take an educated guess based on symptoms, but a fecal identification of the parasite eggs is much better.

Once I get a clean fecal report, I return to my normal preventative treatments. One size does not fit all, and some animals and some breeds are more susceptible to certain worms. Some pastures are harder to control, and the world does not always work the way we want.

No natural remedy will be effective against all types of infestation. Some natural substances will make the gut less hospitable to parasites. But if an overload of parasites is already present, it may not do the trick at all. Proceed with caution and use close observation of the animal during recovery. Seek a veterinarian's advice and get fecal samples tested sooner rather than later. It might be too late, if you wait.

There are products on the market and available at farm supply stores that people have used for many years. Even though I prefer to keep things natural, I will not tell you that you shouldn't use whatever your veterinarian recommends. I will tell you to read the label precautions carefully and follow the instructions for dosing. A good relationship with a veterinarian is an asset to your farm but don't discount the also valuable advice from other goat keepers in your area.

Usually the suggested time frame for treating to control worms is spring through fall. However, the recommendations also state to avoid treating while goats are stressed. The medication itself can be stress inducing. If it seems complicated, it's because a good bit of management must be used when treating or preventing parasites.

The best plan is to get your herd healthy, then, through natural methods, keep them healthy. Good management includes a regular pasture rotation plan for ruminants. There are pre-made natural worm prevention products on the market. For my favorite, see the resource section (page 177).

Many people prefer to treat parasitic worms using a natural anthelmintic like black walnut tincture. If you start your goats on natural preventatives early, you may be able to avoid using a commercial medication.

Black walnut tincture will not cause resistant strains of parasites. It works by making the intestinal tract of the bird or animal inhospitable to parasites. Garlic cloves added to the water and apple cider vinegar will help the intestinal tract maintain a healthy condition, although they will not eliminate parasites.

If you have a black walnut tree nearby, you are ready to make the natural tincture. If not, you can ask around to see if anyone in your area has this tree. Most of the time people are happy to part with the large round walnuts. They are encased in a green hull. The squirrels love to hoard these for winter eating. The nut meat from a black walnut is harder to extract than the more common English walnut. Some have resorted to running the nuts over with a car to crack the sturdy shells! If you can't find the black walnuts locally, some sellers offer small amounts via mail order.

Black Walnut Tincture

MATERIALS

- 10 to 12 ripe, green black walnuts
- 1 quart of vodka (preferably over 50 proof)
- 1-quart jar with a lid
- Small brown glass bottles or other bottles to store the tincture later. Don't forget to label the bottles.

INSTRUCTIONS

1. Working with black walnuts will stain your hands and clothing, so wear gloves and protect your clothing and work surface. Break off the outer hull. If the nuts are fresh, it won't take much pressure to break open the hull. Lightly tap with a hammer if needed.
2. Place the hulls in the quart jar. Pour vodka into the jar until nearly full. You do want to leave some room to gently shake the mixture every day or so.
3. After 6 to 8 weeks, the tincture is ready to strain and bottle. Squeeze or press as much of the tincture out of the hulls as possible. Remember to wear gloves to protect your hands from stains. Discard the hulls in the trash. Adding these to compost isn't recommended. The juglans contained in black walnut can be a growth inhibitor and that is not something you want to add to your compost.
4. Label and date the bottles of black walnut tincture. It will last for years.

Dosage: I used two teaspoons of tincture per 5 gallons of water. Repeat dosage every day for five days in a row, then wait 2 weeks and repeat the dose for another 5 days.

Note* Black walnut tincture can be used for worming chickens, too. The dosage I use is 1 teaspoon per gallon for five days and repeated after two weeks as above.

Natural DIY Worming Mixture

Many herbs and botanicals also contain anthelmintic properties. A loose worm preventative mixture can be produced using dried herbs. This can be fed to each individual goat after mixing the dosage amount with a tablespoon of diatomaceous earth. The recommendation is to use it daily for one week per month during the spring and summer months.

INGREDIENTS

- Fennel seeds
- Pumpkin seeds
- Dehydrated garlic
- Dried oregano
- Dried thyme
- Dried sage
- Dried hyssop
- Dried red clover
- Diatomaceous Earth powder (food grade)

INSTRUCTIONS

1. Mix together equal parts of each ingredient except the diatomaceous earth powder.
2. Use 1 tablespoon and mix with 1 tablespoon of food grade diatomaceous earth powder.
3. Give each goat one dose per day for one week of the month. Repeat during all spring and summer months.

***Disclaimer**

All herbal mixtures are given as suggestions only. No guarantees or scientific results are available for the use of natural and herbal products as worm preventatives or cures. The above ingredients contain health promoting properties. Use all preventatives and suggestions as a part of your management practices as you feel it fits your goat keeping goals.

Hoof Care

All hoofed animals risk foot rot and related foot scald. Both conditions cause soreness and lead to limping around the pasture or kneeling to eat. In severe cases, goats will try to walk on their knees.

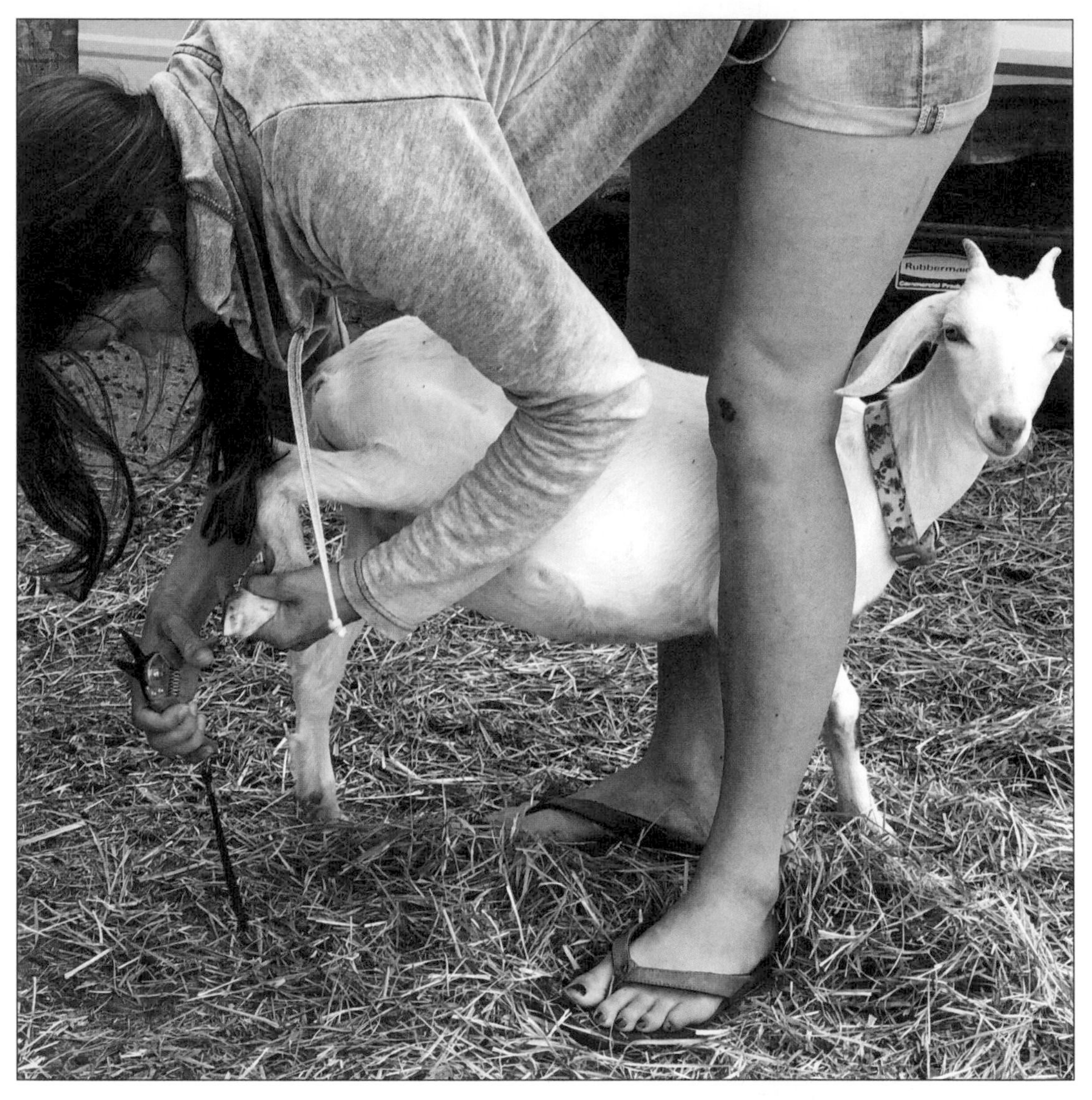

Photo by Victoria Young.

When you see limping or hesitancy in a goat's walk, it's time to take a closer look. Gather the supplies. I keep hoof trimmers, a hoof pick, and a clean rag in the first aid box. Find a calm area of the barnyard and help the goat into the goat stanchion. If you don't have a goat stanchion, enlist the help of someone who can hold the goat still, maybe against the barn wall or fence. Feed some treats to help sweeten the procedure. I have yet to work on my goats' feet without some resistance. Treats and another person make the task much easier.

Wipe the hoof and use the pick to remove any mud accumulated in the hoof. Look for pebbles or debris that may be lodged under a flap of overgrown hoof material. Inspect between the toes. If the goat has scald or rot, this may be painful, so be quick and gentle. An irritated, red area or white and infected-looking tissue are telltale signs of foot scald or hoof rot in goats.

The likelihood of hoof rot in goats has been, in my experience, increased with wet, moist ground and damp weather. Any prolonged periods of moisture can lead to goats limping and holding a leg up. A small irritation or abrasion can let bacteria enter the hoof and soft foot tissue. This can then lead to fungal growth.

Two organisms cause foot rot: *Fusobacterium necrophorum* and *Bacteroides nodusus*. *Fusobacterium necrophorum* lives in the soil. Since it is anaerobic, it needs to grow in the absence of oxygen. This is exactly the situation in deep, muddy pastures or stalls. When the secondary bacterium is introduced, *Bacteroides nodusus* joins with *F. necrophorum* to create an enzyme causing hoof rot.

Cleaning and Treatment of Hoof Problems

Gently clean the affected hoof using a disinfectant solution diluted in water. Be gentle. Keep the goat on dry ground until you have cleaned the stall and brought in dry bedding.

Hoof rot is highly contagious, and easily spread through a herd. Also, clean off the goat stand. Disinfect tools before use on any other hooves.

Check the stall or paddock where the goats are housed. Is the ground damp and moist? Have manure, mud, and dirty bedding accumulated? If so, get that cleaned out and put fresh, dry bedding in the stall. You might find that cleaning more frequently helps reduce the incidence of foot scald and hoof rot in goats. Winter wet weather can contribute to a bad case, and other goats with foot rot can bring the infection to your herd. Always quarantine new stock and check their hooves before adding new goats to your herd.

Treatment

Copper sulfate foot bath is a standard treatment. Pour enough solution into a shallow pan for the goat to immerse affected hooves. You can use plastic pans used for concrete mixing, plastic dishpans, or any large, shallow container. Farm supply retailers sell boots for goats that hold the solution against hooves. Treating foot rot in goats is a lengthy process, but consistency is key to healing.

Foot rot treatments include copper sulfate solutions in easy-to-use spray bottles. Hoof and Heel, a commercially sold product for treating hoof issues, comes in a squirt bottle. This makes it easy to squirt directly onto the sore areas between toes.

Herbal and Essential Oil Treatments for Treating Foot Rot in Goats

Katherine Drovdahl, in her book *The Accessible Pet, Equine and Livestock Herbal*, recommends lavender essential oil and oil of garlic in a blend used to treat foot rot in hoofed animals. Other blends can be made from tea tree oil, cinnamon oil, clove oil, peppermint oil, or sage oil. Note that not all these essential oils are safe for use in pregnant livestock. Ms. Drovdahl's recommendation is 12 total drops of essential oil per tablespoon of olive oil. Never use straight essential oils on an animal or on your own skin. Many essential oils are considered "hot" and can cause irritation similar to a skin burn. Always dilute in a carrier oil, such as olive oil or fractionated coconut oil.

Preventing Hoof Rot

If any animal on the farm property has foot rot, the bacterium will now live in the soil. It will be of extreme importance to keep the ground dry and stalls clean from muddy areas to avoid infection.

Not all limping is a foot rot symptom. Examine the hoof completely before treating. Stone bruising can cause pain and the goat will react by limiting weight on that foot. Arthritis in older goats can lead to lameness and sore joints, and cold weather plays a part in arthritic pain. A goat may hold up a stiff leg after lying down for a long period. If you don't see any evidence of disease or find any tender spots on the bottom of the hoof, check for other causes of lameness. It could be that your goat could use a joint lubricating supplement to combat the effects of aging.

Proper goat hoof trimming won't eliminate the chance of your goat contracting hoof rot, but a healthy hoof is more resistant to bacteria in the environment.

Take these steps to ensure good goat hoof health:

- Trim goat hooves regularly and inspect for signs of injury or disease. Regular hoof trims reduce overgrown areas where wet mud can be trapped.
- Since the bacterium needs moist, anaerobic conditions, keeping stalls clean and dry helps keep it under control. Frequently remove any soaked bedding and muddy, manure-soiled areas.
- Quarantine any new animals joining your herd for at least 30 days. Even your own goats that leave the farm for breed shows or fairs should be quarantined before rejoining the herd.
- Practice good biosecurity on and off your property. Have designated footwear for your goat shelter and do not wear those boots to visit other farms or goat areas.

Unfortunately, once the bacteria that causes foot rot gets onto your property, it is almost impossible to eradicate it. It's worth the extra effort to prevent the bacteria from coming in the first place!

Hoof Trimming

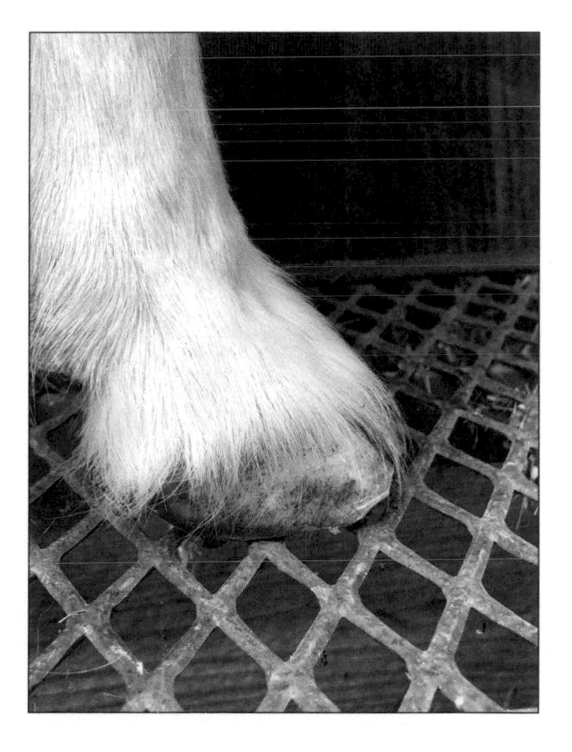

Hoof trimming needs to be tended to every other month. Starting early in a goat's life will help make this less traumatic, but don't be surprised if they still resist. The back feet, especially, seem to be an issue for our goats. Even the older goats do not like having me lift up and hold a back foot for a trimming. I think it is because they can't see me back there, and it probably is a fight or flight response. It helps to have another person stand by their head and distract them with a treat while you trim the back feet.

Putting the goat on a stand makes it easier on the person trimming. I have done a number of hoof trimmings by having someone else hold the animal still, while I trim the hooves. This requires a lot more bending and reaching, but can certainly get the job accomplished. I look at the stand as a great tool to have, but we went many years without owning one, too. Gather all your tools and some treats before you get started.

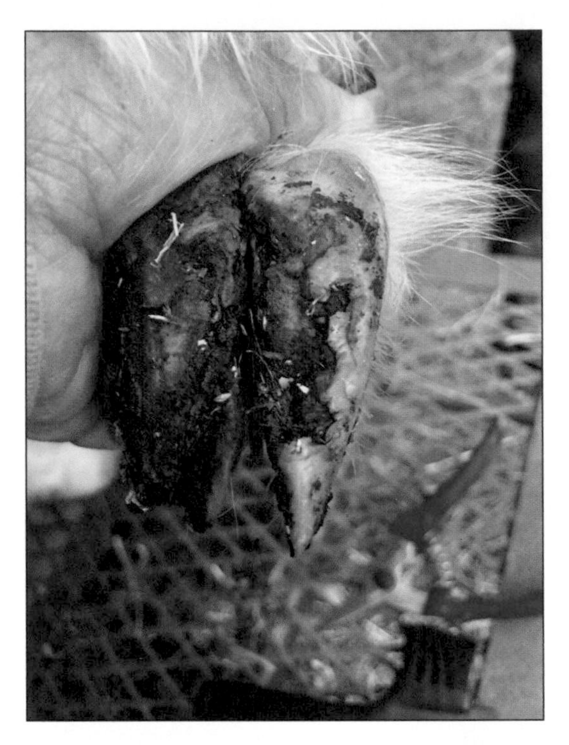

Some of the items I recommend having close by are extra breakaway chain collars, the hoof clippers, yummy treats, an old rag to wipe mud off the hooves, and a sturdy lead rope. Have a plastic container of corn starch ready. If you accidentally trim too close and cause a mild bleeding, applying corn starch will stop the blood flow. Then I apply a dab of antibiotic ointment and it takes care of the mishap. I have never had a serious problem occur after a slight nick of the hoof.

Using hoof clippers makes the job easier because they are shaped to trim hooves. I wear sturdy gloves when doing the hoof trim because the clippers are extremely sharp and animals make sudden moves! I also have used Fiskar's brand garden clippers, but the shape of the blade makes the job a bit trickier.

Keeping up with the hoof trimming makes the job so much easier. It is possible to bring a neglected goat back to some measure of good hoof health, but it takes time and dedication When I have missed a trimming, the amount of overgrowth is amazing. Plan to trim hooves, at the minimum, every other month.

Starting with a front foot, stand to the side of the goat and pick up the leg, bending at the knee joint. The hoof will be facing you. Using the hoof pick, remove compacted dirt from the hoof.

1. Evaluate the side walls of the hoof for separation, cracks, and other irregularities.
2. If the sides have grown over and a flap of hoof is folded under the hoof, gently pry it away from the bottom before beginning to trim.
3. Making your cuts as smooth as possible, trim both sides of the hoof. Remember that each hoof has two sections. The center area is called the frog and is softer tissue.
4. Assess if the front of the hoof needs trimming back also. Usually it does, but if you are good at keeping up with trims, it may not always need to be cut back.
5. Keep the healthy shape of a hoof in mind when trimming. A healthy hoof is wedge shaped and there is no large overgrown part protruding from the front of the hoof.

Goat Horns

Keeping dairy goats, or any breeding stock, ultimately results in baby goats. If the breed is not naturally polled, the kids will have horn buds appear in the first few days after kidding. There are reasons to disbud a kid goat, and reasons that you may choose to not disbud. Disbudding is normally carried out soon after birth. In my opinion, it's one of the harder decisions of goat raising. In any case, it is something to be aware of before you begin a breeding program.

Horned goat. Photo by Victoria Young.

Do You Have to Learn How to Dehorn a Goat?

There is no rule that all goats must be dehorned or disbudded. Some farmers or goat keepers are against the procedure. Some prefer to leave the horns intact as a way for the goats to defend themselves. Horns also naturally aid in cooling the goat. If you choose to dehorn your goats, the disbudding needs to be performed in the first few days, preferably day 3 to 5, after birth.

Reasons to consider dehorning goats include:

1. Having small children on the farm and safety issues of goats with large horns around children.
2. Horns can get caught on fences, feeders, and other things, sometimes leading to injury or death of the goat.
3. The doe can injure the kids with her own horns.
4. The goats can injure each other while playing or battling for dominance.
5. The horns can injure you while you're milking or performing other routine care.
6. The breed standard requires dehorning/disbudding for registration or participation in breed shows.
7. Another consideration when deciding if you will dehorn kids or not is what you already have in your herd. Horns are one way a goat defends itself. And if you have ever been on the wrong side of a goat butting you, then you know it can really hurt. Horned goats head butting polled goats isn't a fair match and can result in injury. Not all horned goats are aggressive and prone to fighting, but most goat herds I have seen, including my own, have a fair amount of head butting and jostling for position.

With any invasive procedure carried out with livestock, we need to be prepared for unhappy outcomes. While it is rare to lose a goat kid to disbudding, it can happen. Leaving the dehorning iron on the horn buds for too long or pressing too deeply can result in brain damage, infection, and death.

If you feel comfortable after learning how to dehorn a goat, you will be ready to take care of the kids in your herd. If you cannot get past the thought of performing the disbudding, maybe another goat breeder will take care of it for a fee.

Disbudding Box

Project and photo by Lesa Wilkes, Bramblestone Farm

The disbudding box is a wooden box, not much bigger than the goat kid. The kid fits snugly into the box and the head is placed through the cut-out opening. A lid is closed, leaving only the head protruding. The box holds the kid securely for disbudding and tattooing or tagging the ears. It is possible, although probably not the best idea, to have one person hold the kid tightly while the other person burns the horn buds with the disbudding iron. Whenever possible, use the disbudding box.

MATERIALS

- From a sheet of plywood cut the following pieces to construct the disbudding box.
  - Belly Board—1"x5"x3½" (#1 in Sketch). Helps prevent the kid from trying to lay down in the box. Position about 7" from the front of box.
  - 2 End pieces—1"x5"x15¾"
  - 2 Side pieces—¼"x16"x24"
  - 1 Top piece—1"x5½"x24"
  - 1 Bottom piece—1"x5"x24"
- Additional hardware
- Neck piece—optional but helpful
- Latch for holding the top closed
- 2 hinges for the top
- 1 handle for carrying

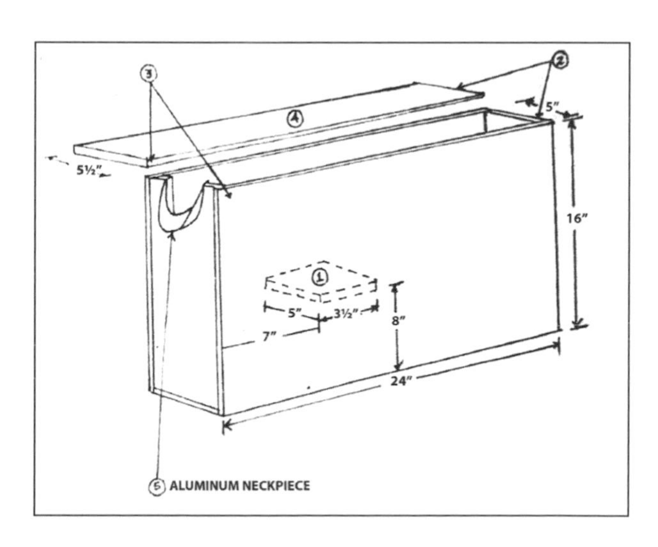

INSTRUCTIONS

1. Cut out the neck opening as shown in the plan.
2. Attach the front and back panels to the bottom panel.
3. Attach the sides.
4. Attach the hinges to the top and then attach to one side of the box.
5. Attach the closing latch to the opposite side from the hinges.
6. Place the belly board inside the box and secure with screws to the sides.

Disbudding Procedure

The disbudding iron is an electric tool with a handle and long metal rod that becomes extremely hot. The enclosed end of the metal rod is held against the horn bud, long enough to cauterize the blood supply and stop any horn growth, but not for too long, or infection or brain damage can occur.

MATERIALS

- Ice bag
- Tetanus antitoxin injection
- First aid kit
- Disbudding iron

INSTRUCTIONS

1. Apply the ice bag first to numb the area.
2. Inject the kid with the Tetanus antitoxin before beginning the disbudding procedure. A well-stocked farm first aid kit, containing an antibacterial salve, gauze, and other products should always be close at hand when working with animals.
3. Test the heat of the iron on a scrap of lumber. It should leave a burnt ring within two seconds.
4. While kid is secure in the disbudding box, locate the horn buds. Apply the disbudding iron to the horn bud. Apply light pressure to ensure that the iron

cauterizes the area. Hold for 3-4 seconds. The result should be a copper colored ring. For male goats enlarge the area on the next attempt by making an overlapping circle with the first. This is to ensure you get the entire horn bud area and the scent area.
5. Repeat procedure on the other horn bud.
6. Repeat the procedure on both horn buds for two seconds.
7. Post-disbudding care includes keeping an eye on the horn buds for any signs of infection or bleeding. As the scab prepares to drop off, minimal bleeding might be seen. Any heavy secretions or drainage should be treated by a veterinarian. While antibiotics are not routinely used for the disbudding procedure, having a farm first aid kit stocked with a good quality antibacterial spray is always a good idea.

Before and after photos of disbudding.

What Are Scurs?

Scurs are smaller, misshapen horns that grow if some of the horn bud is not destroyed in the disbudding process.

Polled vs Horned

Horns are the result of genetics and the traits carried by the genetic code. Polled goats carry the trait to create polled kids. It gets complicated, though, if you have a polled doe that is actually a heterozygous genetic carrier, meaning she carries the trait for both polled and horned offspring. Delving even deeper into polled genetics, it's important to note that the horned genetic trait is lower than the polled trait. The polled gene is considered dominant. Therefore, if you have a true (homozygous) polled goat mated to a true horned goat, all the kids will be polled (without horns). However, they will carry the genetic trait for horns that could result in horned kids if they were mated with a horned goat. Simple, right?

Oh, one more fact. Breeding polled goats to other polled goats can result in hermaphrodite offspring.

Neutering Male Goats

Controlling the population on our backyard farms and homesteads is of primary importance. All animals require care, feed, veterinary services, and room to move around. Overpopulation can occur quickly.

If you are not planning to breed your livestock, the most responsible course of action is to castrate or neuter your livestock. Animal behavior is one reason to go this route. Males that are not castrated tend to be more aggressive, which can be a problem if you are not prepared to deal with this trait. Food aggression can be annoying and dangerous. An eighty-pound buck jumping on you to get the food bucket can result in you being injured.

Unwanted babies is another reason to neuter livestock on your homestead. Males left in the same field with the females often become territorial. Do not think that you can wait until you see mating behavior before separating the animals. By the time you actually witness mating behavior, you are probably expecting baby animals.

There are multiple methods to use to neuter livestock. Surgical castration, Burdizzo method, and banding with an elastrator are common methods used by farmers. Burdizzo and the elastrator for banding are both bloodless methods. Using a bloodless method results in less likelihood of attracting flies and subsequent flystrike.

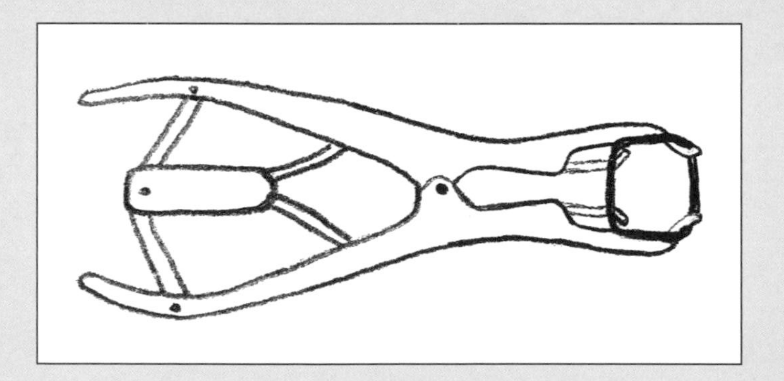

Rumen Health

All goats are ruminants. Goats can be kept for different purposes, but no matter the goat breed, all goats digest forage and plant material in a multi chambered set of stomachs. As the goats browse and begin to ingest plants, the material travels to the first stop, the rumen. The rumen is continuously moving three layers of healthy content, gasses, newly eaten roughage, and a bottom layer of slurry-like grain and saturated roughage from the previous day. In the rumen, millions upon millions of bacteria, protozoa, yeast, and fungi are present to get to work. The reticulum is closely connected to the rumen and grows and supplies the rumen with bacteria to start digestion.

Chewing cud is a curious term describing the second chewing of food that has been floating in the rumen. Goats spend the majority of their day either eating or chewing cud. The cud is forced back up the esophagus and into the mouth where it is chewed again. This process results in additional breakdown of the food, along with large amounts of saliva being sent on to the rumen. The saliva helps neutralize the stomach acid, keeping the microbial life happy.

The food is forced from the rumen during the regular contractions and it continues its journey through the reticulum, omasum, and abomasum. Ruminants are efficient processors of fiber foods. When the process is complete, the foraged food has been broken down into usable sugars, starches, and amino acids from the plant protein content.

It's all a delicate balance, and a large amount of gas is produced and expelled every day by ruminants. The gas has to be relieved by burping or rectal passing. Any change in diet can cause disruption in the rumen activity, causing gas to back up, leading to bloat. Bloat can quickly become life threatening in ruminants.

Watching a goat suffer the effects of bloat is scary. The goat that was healthy in the morning can be near death by evening, or even sooner. I realize that might sound dramatic, but bloat is a life-threatening condition.

Bloat is more likely to strike goats in the spring when they are not used to lush pasture and growth.

Conditions that are likely to cause bloat:

1. Pastures that have a high legume content
2. Lush pasture that is still wet from dew
3. Alfalfa pasture
4. Overeating grain, or another animal's feed (keep those feed cans covered and secured!)

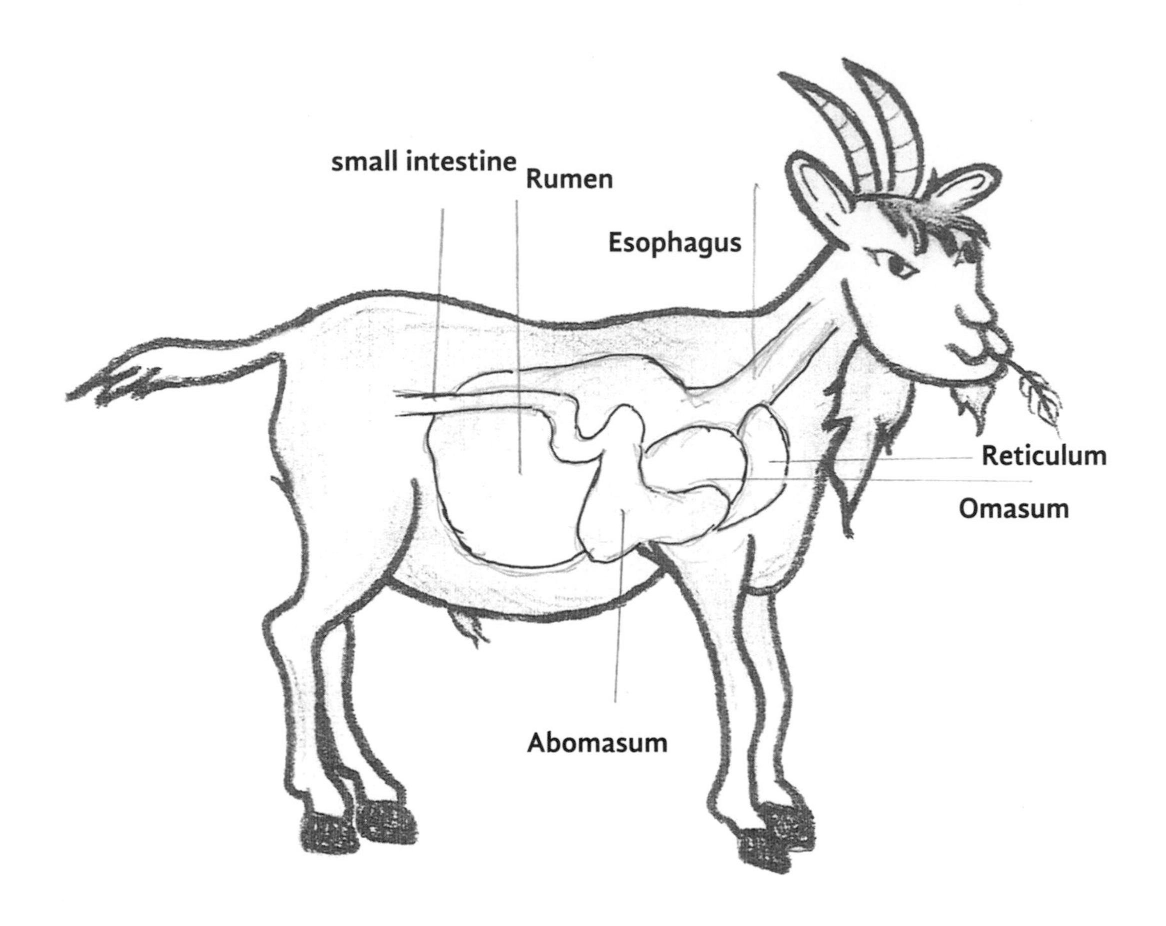

One early sign to look for is a larger than normal protruding left side. The goat will act uncomfortable, perhaps nipping at its side, and may appear glassy eyed. It will not be eating, chewing its cud or belching.

To help avoid bloat, reduce the exposure to fresh lush pasture first thing in the morning. One way we limit pasture is to keep the goats in the barn during the night. They have access to hay and water, and we replenish the hay first thing in the morning. When they are released to forage, they are not ravenous with the morning hunger, and the dew has dried on the plants and grass. We gradually lengthen the time on pasture as their digestive system gets used to the new food. This has worked very well for us and has also lowered the incidence of intestinal parasites.

Bloat Remedy

Warm water, olive oil, and baking soda are commonly given to relieve the bloat and enable the goat to belch. You will need a drenching gun on hand to get the best result when trying to get this remedy into your goat.

INGREDIENTS

- 1½ cups warm water
- ½ cup olive oil (mineral oil or canola oil can be substituted)
- 2 ounces baking soda dissolved in the liquid

INSTRUCTIONS

1. In a mason jar, mix the ingredients.
2. Fill a drenching syringe with the mixture.
3. Insert the drenching syringe into the goat's mouth. You may need to insert your thumb in as the goat may be clenching its teeth.
4. As you begin to release the liquid, make sure the goat is swallowing.
5. If you can, continue to walk the goat, to help move the gas out. Notice when the goat begins to belch.

The oil helps to break the surface tension in the rumen and allow the water and baking soda to mix in and cause a burp.

Anemia in Goats

Anemia in goats is a symptom that can point to internal parasites. One internal parasite, the barber pole worm, *Haemonchus contortus*, can be devastating to goats and other ruminants. This parasite has a sharp tooth that slices into the stomach (abomasum) and makes a meal of the blood. The live worm completes its journey there and lays up to five thousand eggs. These eggs are excreted within the goat feces pellets, hit the ground, and are released into the dirt. The larvae cling to the moisture of plants and are ingested by the goat as it grazes.

Barber pole worm thrives on wet, warm pasture. Hot, dry weather and extreme cold break its life cycle or cause it to lie dormant. The worm can even lie dormant in the stomach of the host, which is one reason it is such a problem parasite to deal with.

Symptoms of infestation include thin coat and dry brittle hair, bottle jaw, a protruding, lower jaw swelling, acute anemia, listlessness, and death. The animal can decline rapidly.

It's important with parasite control to not overtreat with dewormers and thus add to the resistance. A simple fecal exam will give you the information necessary to treat for the correct parasite. When dealing with barber pole worm it is vital that you work quickly to begin treatment.

Anemia in goats is life threatening. If you are not sure of the cause of the anemia, contact your veterinarian for assistance.

FAMACHA Score

The FAMACHA system was developed in South Africa to evaluate the level of anemia using a range of color showing in the eye mucus membranes. The scoring is based on five levels of color shown in the lower eye mucus membrane, when the observation is performed according to specific directions. Level 1 corresponds to no anemia and level 5 is severe anemia. It was developed specifically to test the level of anemia from the barber pole parasite using a special score card and training. Certification and scorecards can be obtained by visiting this website from the University of Rhode Island and watching the training videos: https://web.uri.edu/sheepngoat/.

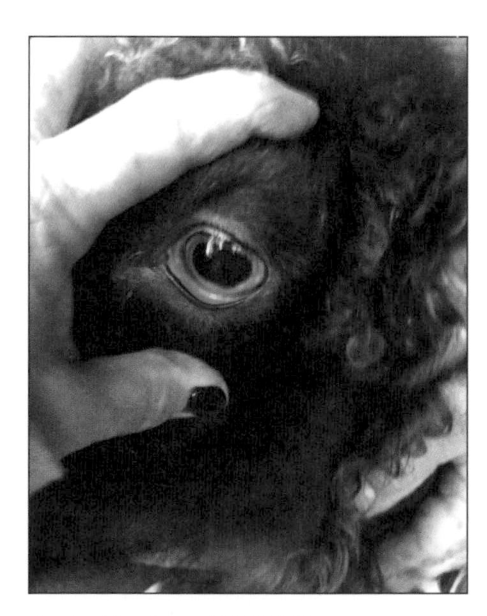

Goat with a low FAMACHA score.

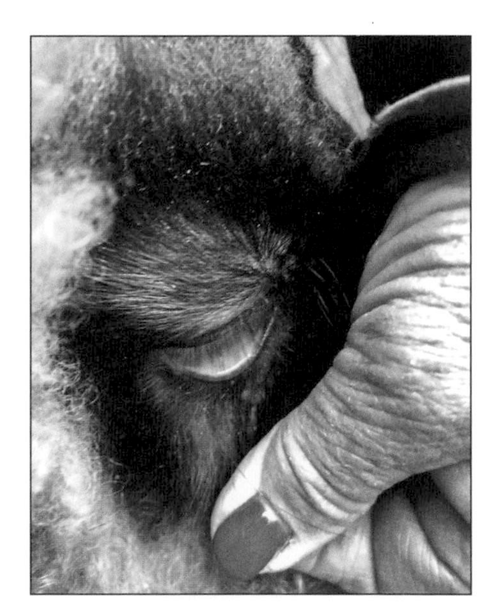

Goat with acceptable FAMACHA score.

Urinary Calculi

When mineral crystals develop in the urinary tract, the condition is called urinary calculi. If the crystals block the underdeveloped or smaller urinary tract, a blockage occurs. The condition is painful and usually fatal.

Early castration can lead to future urinary calculi blockage. To lessen the chances in your male goats, postpone castration until four months of age. Be aware that the goat will most likely be fertile and attempting to mate the females, including his own mother, before this point. Separate the unneutered males or risk an unwanted pregnancy.

Vaccinations and Diseases

Chlamydia Abortion—The vaccine to prevent chlamydia is given four to six weeks after a successful breeding is confirmed. This is a deadly disease for goats and humans. Making it part of your health routine will prevent late term abortions in most cases and protect you while you assist a doe giving birth. Chlamydia is easily spread through the herd.

Cornybacterium pseudo-tuberculosis—This vaccine is referred to as CLA and is given to kids at six months and three weeks followed by annual boosters.

Enterotoxemia and Tetanus—The CDT vaccine is given to kids between one and two months old with a follow up one month later. Does should be revaccinated four to six weeks before kidding and all goats should receive an annual booster shot. The 3-way vaccine protects against three forms of the clostridium bacteria which can cause tetanus and overeating disease. Clostridium C and D, *clostridium perfringins* type C and D, and the tetanus toxoid combo vaccine is given at four weeks and again at eight weeks. Even if you give no other vaccines, this should be considered as the one to administer. It is available through your livestock veterinarian and in some livestock supply stores.

Enterotoxemia is also referred to as overeating disease. Goats will show signs of bloat, pain, and fever. There is no treatment once this has been contracted.

Tetanus, *clostridium tetani,* results from a tetanus bacterium entering through a wound. It creates stiffness and muscle spasms leading eventually to death if not treated. The vaccine for this is included with the CDT vaccine and given at one month of age, with a follow-up vaccine one month later.

Orf or Soremouth—This vaccine should be given to all goats annually. Soremouth presents as mouth scabs and lip sores lasting from one to four weeks. You may also see sores around the nostrils, eyes, vulva, and udders. Soremouth spreads quickly and humans are at risk of contracting the virus. If you keep a closed herd you may not encounter this. Goat owners who show goats are at greater risk and should consider the vaccine with their veterinarian's input.

Pasturella Pneumonia—This vaccine is recommended for kids and newly acquired stock. Vaccine is given in two doses four weeks apart and repeated annually.

Rabies—Veterinarians recommend a rabies vaccine to protect livestock at risk of being bitten by wildlife carrying active rabies infections. The vaccine is administered annually. Symptoms in goats range from loss of appetite, frothing at the mouth, aggressive behavior paralysis, and other changes in behavior.

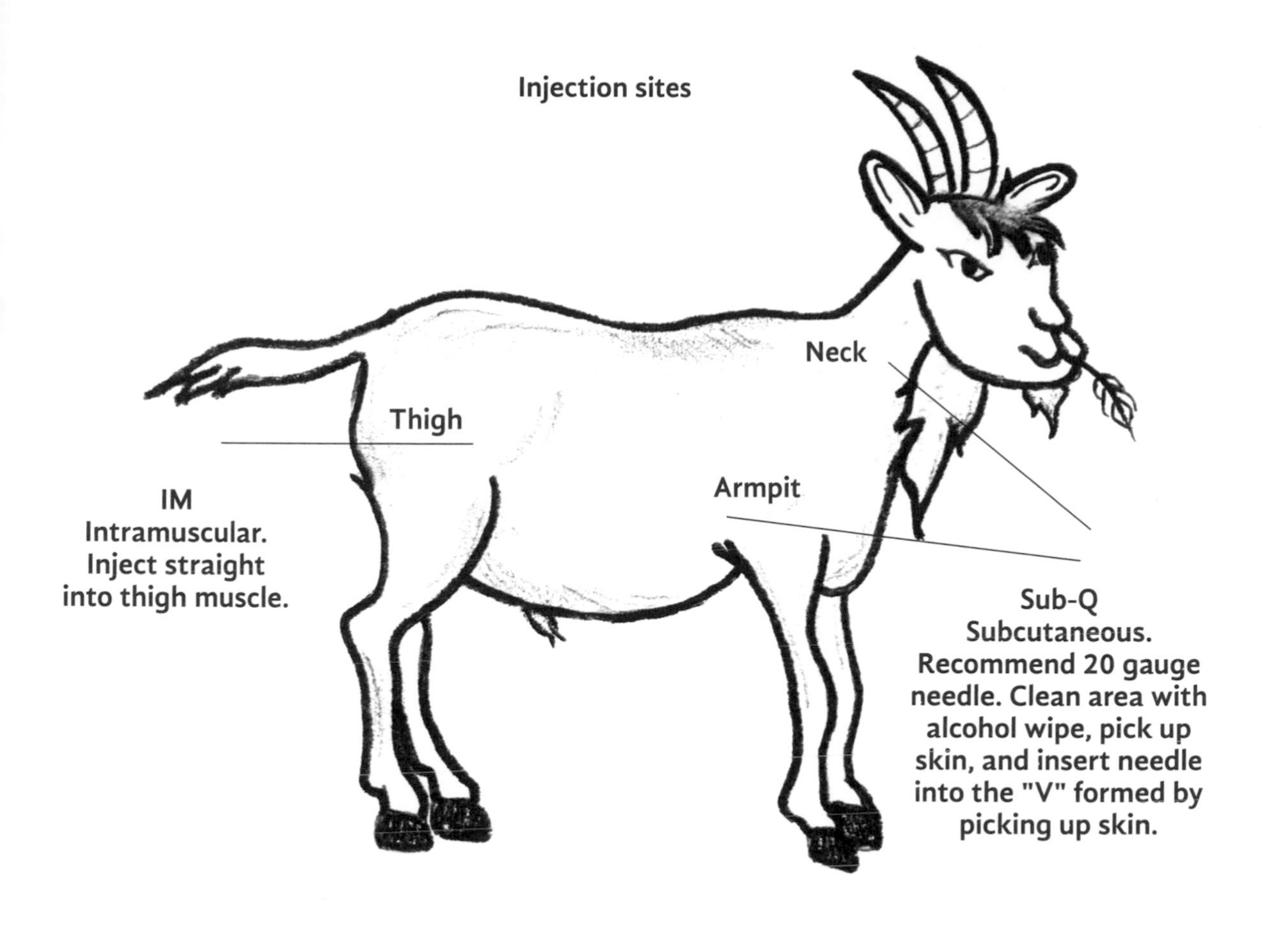

Electrolyte Formula

When goats are sick or injured, they can quickly become dehydrated. Some likely causes of dehydration are scours, fever, extreme weather, wounds, or other injuries. Making an electrolyte solution at home is quick and easy. You likely have the ingredients in your pantry.

INGREDIENTS

- 1 teaspoon baking soda
- 2 teaspoons salt
- 8 tablespoons molasses or raw honey
- 1 gallon of warm water

INSTRUCTIONS

1. Mix all ingredients together.
2. Fill a large syringe with the solution.
3. If you are handling the goat by yourself and the goat is still standing, straddle the goat. Lift the goat's head slightly using one hand under the chin. Insert the syringe towards the back of the mouth and slowly dispense the electrolyte solution into the goat's mouth.
4. Do not administer the electrolytes to a goat that is down and unresponsive. Call your vet.

Flystrike

Farming and flies are no surprise. If you have livestock. you most likely battle the common stable fly, too. If animal manure piles up and is not composted properly, your fly population will increase past the point of annoyance. This can be a real danger for your goats because with the right conditions, you may encounter flystrike.

Diarrhea, loose stools, manure stuck on the anal opening, and wet fur all attract flies to your goat, which can lead to flystrike. Flystrike can happen to any animal. While it is linked to unsanitary conditions, it does not always mean that you aren't taking proper care of the animal. Just a few hours of an animal having caked on manure or runny poo stuck to its fur can be enough to attract flies and lead to flystrike. If you are raising Angora or Cashmere goats, the hair might hide the beginning of flystrike.

When watery feces or matted fur with feces occurs, the flies lay their eggs on this area. This becomes the perfect breeding ground for flies. Fly eggs mature and hatch quickly, which is a major factor in flystrike. Not noticing the presence of diarrhea, wet fur, or urine-soaked fur for even half a day can give the flies time to lay thousands of eggs. Before you know it, flystrike has started. The maggots will continue to eat the flesh and internal organs of the animal. Death can occur quite quickly if not noticed and treated.

Flystrike Treatment

1. Clean the wound. Manually remove all maggots. The wound may be a deep hole or pocket created by the larvae and maggots. It must be cleaned out completely and treated daily.
2. Confine the goat to an area where you can monitor the progress and easily treat the flystrike. Isolate the animal if others in the herd are pestering the wound. Perform daily wound care by cleaning out the wound with sterile saline solution. Wash out the area with an antibacterial soap. The affected area may be tender, so handle the wound as gently as possible. Gently dry the affected area.
3. Apply a triple antibiotic cream (one that does not contain a pain reliever) inside and outside of the wound.

Keep the animal in a dry, well-ventilated area. If there are still loose bowel movements, treat this also. It is important to keep the feces from sticking to the animal's genital area. Flystrike treatment will also include daily cleaning and removal of any manure and feces from the area, in order to not attract more flies. In livestock, using a fly repellent cream such as Swat around the affected area will also deter more flies from trying to attack the wound.

*Flystrike can result in death. If you feel that the treatment is not progressing well, call a veterinarian immediately. There is always the chance that the open wound has developed an infection.

First Aid Kit for Goats

If you have goats, you need a well-stocked first aid kit. Goats are mischievous by nature, which often leads to the need for first aid. Include items for treating external wounds like cuts, bruises, and sores. Goats may need internal first aid too, particularly when internal parasites cause a problem.

One of the first things you may notice after acquiring goats is the lack of available emergency veterinarian care. In some areas you may not be able to have your ailing goat seen the same day as an illness or accident occurs. Your vet might give

HOOF 'n HEEL
CVS
SAFE GUARD
NITRILE POWDER FREE GLOVES

you advice over the phone in order to assist the animal. Having a well-stocked first aid tote ready to use may save your goat's life.

Bloat is a common ailment with goats, especially if they are put on new forage or pasture. Bloat can be remedied if discovered early. Keeping simple baking soda on hand in the goat medicine cabinet saves time and might save the goat's life. Read information about bloat on page 77.

Baking soda offered free choice allows the goat to self-regulate the pH of the rumen. Keep **olive oil** on hand for mixing up a bloat remedy solution (see page 79). The oil breaks the surface tension of the bubbles of gas that are trapped in the rumen.

Goats decline rapidly when sick. Sometimes the only chance to save the goat's life is being prepared with a good first aid box.

We add the following items into our goat first aid box. These are items we purchase from a livestock supply retailer. Some of the items can be purchased at your local drug store.

A **digital rectal thermometer** should be in any farm first aid box. You do not need to purchase a specific livestock thermometer, although the attached string at the end of the livestock thermometer is a good idea. Thermometers have a way of getting sucked into the rectum and large intestine if you aren't holding on to them. The first thing a veterinarian will ask you over the phone is if the goat has a fever. A normal goat temperature reading should be between 102 and 103°F. Being ready with this information saves time and allows the vet to suggest treatments based on symptoms. A good pair of **scissors** and **tweezers** are good additions to any medical kit.

For eye injuries, our choice is **Terramycin Ophthalmic Ointment**, purchased over the counter from livestock supply retailers. This, along with **Vetericyn Ophthalmic** ointment, are the first line of defense for an eye infection or injury in our goat herd.

Along with the goat's natural energy comes unwanted cuts, scrapes, and injuries. **Anti-bacterial sprays** are a good first line of defense when a wound occurs. I use an inexpensive bottle of **contact lens saline solution** for flushing out the wound. **Hydrogen peroxide** and **betadine solution** are also kept for wound care. A bottle of **rubbing alcohol** is useful for cleaning up the scissors, tweezers, or other non-disposable instruments used. My favorite antibacterial antifungal spray for wounds is **Veterycin Spray**, sold in livestock supply stores, pet stores, and online retailers.

Don't forget the **bandages**, **antibacterial salves or ointment**, and a good supply of **gauze pads (4 x 4 and 2 x 2 size)** Include a box of **human bandaids** too. **Vet Wrap/Cohesive bandage** is used for keeping the gauze or cotton bandages in place. Goats try to eat the bandage soon after you apply it. If the weather is wet, a strip of **electrical tape** resists moisture best. I will add it to the final vet wrap layer to hold the bandages in place.

Round out the supplies with a few sizes of syringes, drenching guns, B vitamin complex, and activated charcoal.

Your kitchen cabinet has products for the first aid box too. **Cornstarch** is good for slowing blood flow. I have used it when I cut too closely on a hoof trim or nicked the skin during shearing on our fiber goats. **Tea bags** soaked in warm water can also stop or slow blood flow. If you grow **yarrow** in the herb garden, chop a handful and apply to the bleeding area. Yarrow is a good plant for slowing blood flow and **Epsom salt** is a good aid for soaking bruises on legs and feet.

Lubricant, **paper towels**, and **disposable exam gloves** are included in our goat medicine cabinet. There will be times you are glad you have them!

Suggested Items for First Aid Box

- Antibacterial wound spray
- Baking soda
- Bandages
- B-Complex Vitamin (injectable)
- Betadine solution
- Cohesive bandages (Vet Wrap)
- Cornstarch (can be used as a blood stop)
- Digital thermometer or livestock thermometer
- Epson salt
- Exam gloves
- Gauze pads
- Hydrogen peroxide
- Obstetrical lubricant
- Olive oil
- Paper towels, or clean rags
- Saline solution
- Salves and ointments
- Scissors
- Syringes and needles (20–22 gauge)
- Tea bags (can be used as a blood stop when wet)
- Terramycin ophthalmic ointment or Veterycin Ophthalmic
- Tweezers

First Aid Carry Box

We store our first aid supplies in a plastic tote box in the feed room. Often we only need a few of the items in the box and it's quite heavy. To take just a few items to the stall to treat a wound or other first aid issue, we use this simple carry box.

MATERIALS

- 2 pieces of lumber, 1'x6" for the sides
- 1 piece of lumber, 1'x6" for the base
- 2 pieces of lumber, either 6"x6" or 6"x9" if you want the ends to be higher, as shown in the photo.
- 1 piece of web strapping 1 yard in length (adjust length as desired)
- Saw
- Hammer and nails or screws and drill driver

INSTRUCTIONS

1. Attach the sides to the base piece of lumber
2. Attach the ends. Taper the ends if desired as shown in the photo. This will not affect functionality of the box
3. Nail, screw, or staple the strapping to the end pieces for a carry handle. Adjust to your liking. Use industrial staples only if desired to attach in that manner.

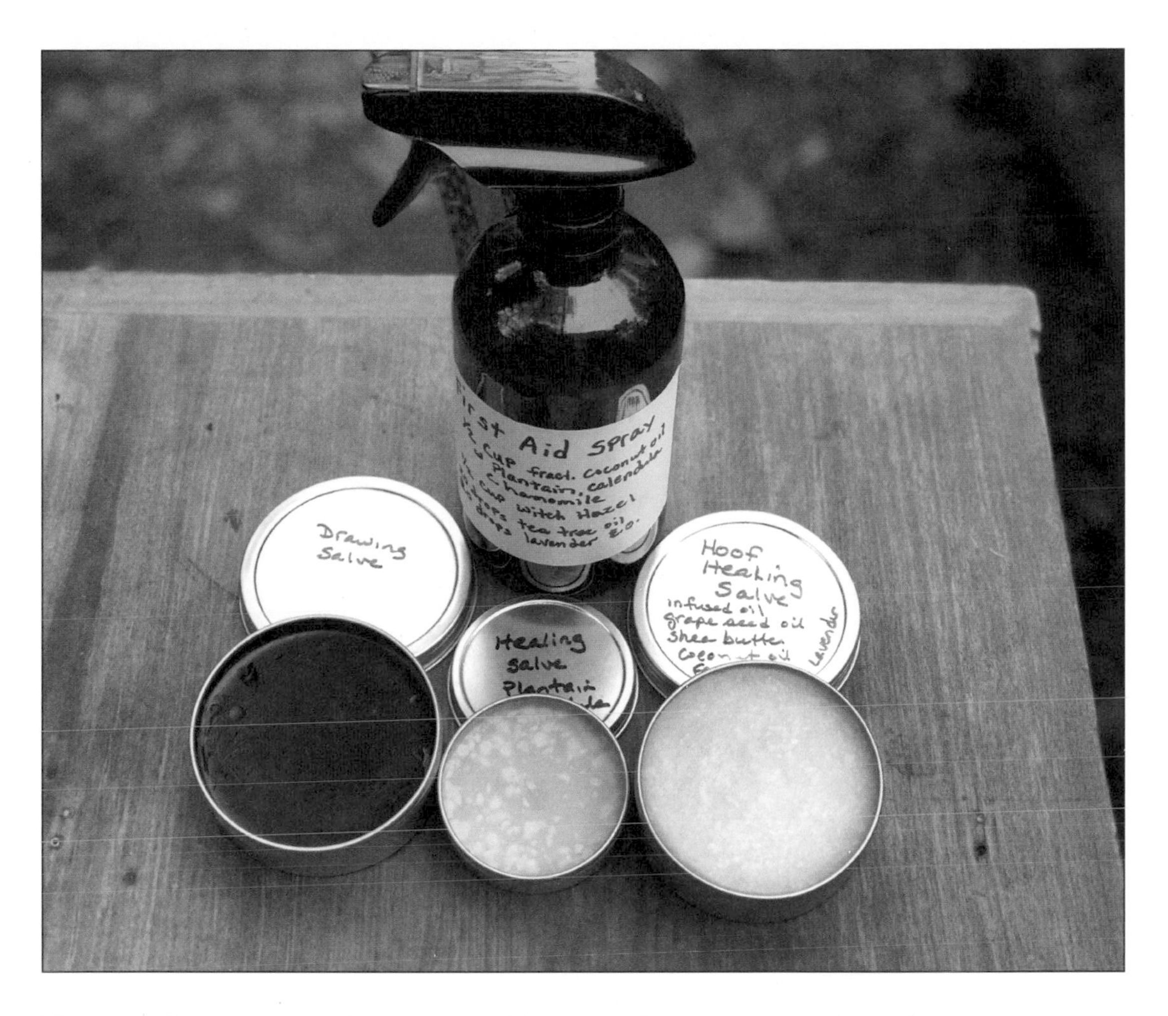

Topical Sprays and Salves for Wounds, Injuries, and Skin Issues

When your goats have an injury, sore, or wound, they are going to think you are a rock star when you whip out these all natural, healing remedies. Herbal and natural first aid is simple. Often, your handmade salves will resolve the problem and everything will return to normal. I am not advocating for leaving the commercial products out of your first aid kit. Quite the contrary. First aid has many levels. I like to start with my natural salves and sprays. If I see that an injury is not improving, then I reach for commercial products.

Drawing Salve

INGREDIENTS

- 6 tablespoons of infused oil (see page 156, plantain and calendula are my choice)
- 10 grams beeswax pellets
- 15 grams bentonite clay
- 10 grams activated charcoal
- 30 drops lavender essential oil
- 15 drops tea tree essential oil

INSTRUCTIONS

1. Combine the infused oil and beeswax in a pint mason jar. Set the jar in a pan of simmering water to melt the beeswax.
2. When the wax and oil are melted together, remove the jar from the hot water bath.
3. Stir in the clay, charcoal, and essential oils.
4. Immediately pour the mixture into 4-ounce containers. The 4-ounce canning jars are good for this purpose, or purchase a 4-ounce tin container.

Antiseptic Spray

Antiseptic sprays are handy for many barnyard applications. Making your own solution is very simple but requires a little bit of prep work.

Plantain infused oil is packed with healing properties. Plantain can be used as a poultice right on the spot by chewing up a few leaves and packing a wound with the mush. If that's not your idea of a good day, using an antiseptic healing spray is the next best start to healing a wound on your goat. (You can use this spray for you and your family too!)

Pick some fresh plantain leaves and allow them to dry over 24 to 48 hours. Place the dried leaves in a pint mason jar and cover with fractionated coconut oil (coconut oil that remains in a liquid form and does not harden). Allow the dried plantain to infuse the oil using the quick method described in the box on page 156.

MATERIALS

- 1 spray bottle (preferably amber glass spray bottle)
- ½ cup fractionated coconut oil or olive oil infused with dried plantain, calendula, and chamomile (see page 156)
- 20 drops tea tree essential oil
- 20 drops lavender essential oil
- ½ cup witch hazel (alcohol-free type preferred)

INSTRUCTIONS

1. Add all oils and witch hazel to the glass bottle. Shake to mix.
2. Before each use, shake gently to re-mix the oils. Spray on wounds, scrapes, and bug bites as needed for a first aid spray. If redness or irritation continue or worsen, contact your veterinarian.

Antiseptic Salve/Ointment

Much like an antiseptic spray, a salve is a first line of treatment for scrapes and wounds. The difference is staying power and how the animal reacts to having the area touched. Some animals will shy away from a wound being treated. Use either a salve or spray to clean and begin treatment.

MATERIALS

- Pint mason jar
- 15 grams (½ ounce) coconut oil
- 15 grams (½ ounce) beeswax
- 4 ounces infused olive oil with any healing herbs or combination: dandelion, plantain, chamomile, oregano, wild violet, calendula (see instructions on making an infused oil on page 156)
- 20 drops tea tree essential oil
- 5 drops Vitamin E (promotes healing and keep oils from becoming rancid)

INSTRUCTIONS

1. In a glass jar, melt the solid oil and the beeswax, using one of the double boiler methods described above.
2. Add the infused oil and warm.
3. Remove from the heat and add the tea tree essential oil and the Vitamin E.

Hoof Salve for Cracked or Damaged Hooves

Goats don't have as much trouble with cracks in the hooves as horses tend to get. But occasionally the hoof will show cracks, especially in dry weather.

MATERIALS

- 15 grams shea butter
- 16 grams raw beeswax
- 15 grams coconut oil
- 2 ounces grapeseed oil
- 2 ounces olive oil infused with calendula (see page 156)
- 15 drops frankincense essential oil

- 15 drops lavender essential oil
- 10 drops lemon essential oil
- 5 drops helichrysum essential oil

INSTRUCTIONS

1. Melt the shea butter and the beeswax together in a pint mason jar, using the double boiler method. Add the liquid oils and stir to mix. Pour into shallow containers such as 4-ounce canning jars or small tins.
2. Store in a cool place to avoid having the salve melt. Smear a small dab of the salve on the hard surface of the hoof and let it soak in. Of course, it will soon be covered with dirt and bedding, but the salve will still do its work. Wipe the hoof to clean before reapplying more salve.

Chapter 4: Goat Activities

Why Play Structures Are Important to Goat Keeping

Goats are playful beings. They jump, romp, and bring joy to the barnyard. When the goats have no outlet for their inexhaustible energy, they quickly turn to destructive, or escapism, behavior. While goats on range land may not exhibit these behaviors, goats kept in a barnyard setting or smaller yard will do better with some fun additions to their paddock.

As with any backyard, small farm, or homestead project, the cost does not have to be a contributing factor to enriching your goats' lives. Think reuse, recycle, and reduce! A brief survey of our property and business yard revealed many possible goat enrichment projects. Tires, pallets, scrap lumber, and tree stumps all were reused for this new purpose. Elaborate structures are great too, if you enjoy that type of project. The following photos show you various ways that your goats can be kept busy. This is the fun part of goat keeping. The goats enjoyed the minimal construction projects as they investigated and stuck their noses into every project.

Safety Concerns

In order to keep your goat yard as safe as possible, you will have to think like a goat. Look for areas where horns might get caught. Are there possible obstacles that the goat could jump onto in order to climb over the fence? What about protruding objects at goat eye level? It's not easy to switch gears and think like a goat, but the longer you observe your herd in their environment, the possible hazards will reveal themselves.

This half buried tire is a simple project that provides a climbing structure and play opportunity for kids.

Here is a play yard made with pallets and tires and ramps. Provides multiple places for climbing, jumping, and goat games.

Photos by Ann Accetta-Scott, *A Farm Girl in the Making*. If your property has natural features, incorporate them into the goat play yard. Goats will happily climb and play on just about anything so use what you have readily available. We often put some grain on the top of a tree stump to encourage their natural curiosity. Electrical spools can be acquired from contractors, often for free.

Photo by Elizabeth Taffet.

Tree stumps in the play yard can serve the purpose of a feeding area.

Simple One-Step Project for Itchy Goats

MATERIALS

- A used broom head such as those used for a push broom
- 2 (4") screws and drill driver

INSTRUCTIONS

1. Screw through the brush, attaching it to the fence post or barn wall in two places. Goats will rub on the bristles when removing their undercoat in spring or just to scratch an itch.

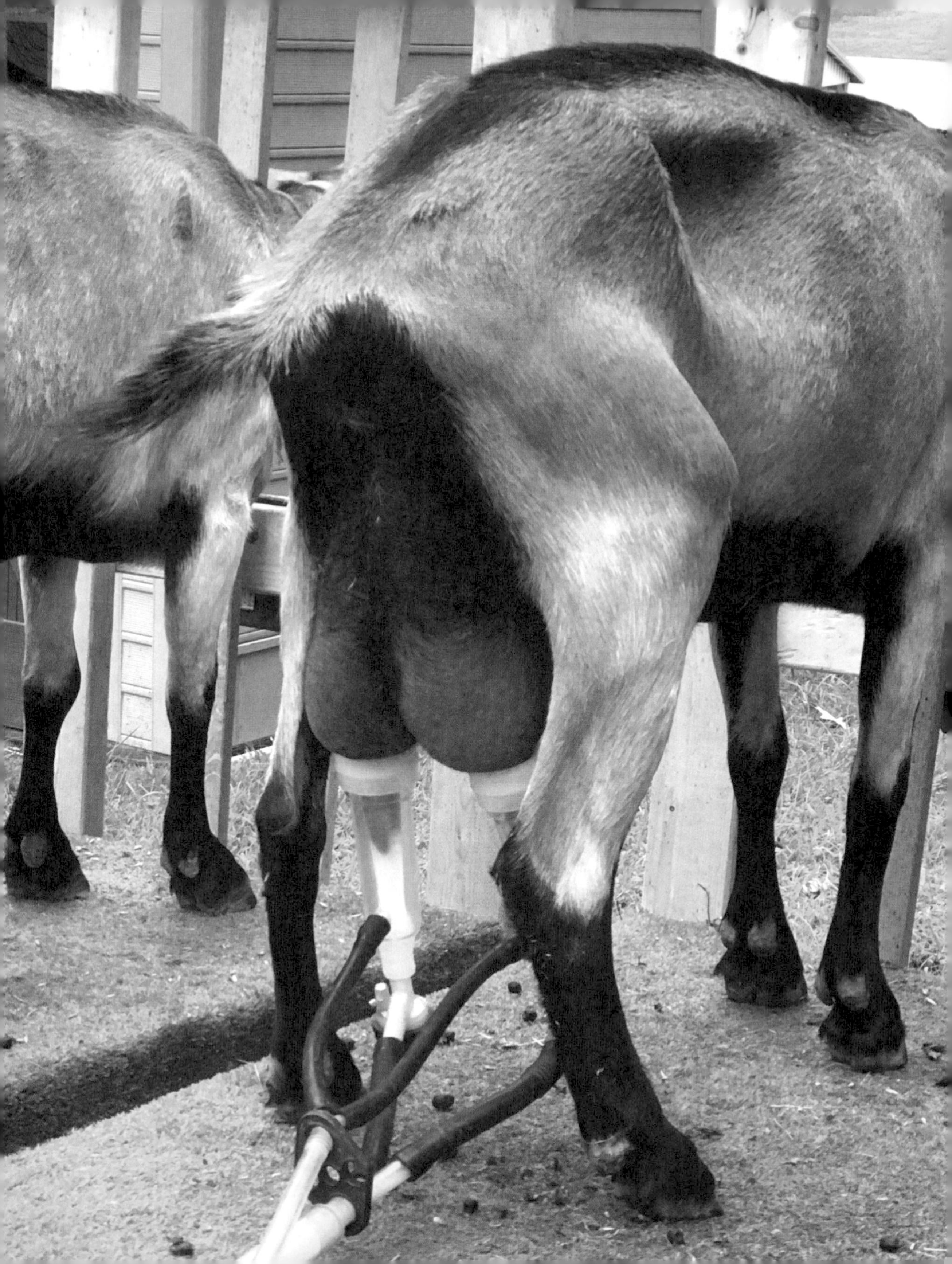

CHAPTER 5: Dairy Goats

Dairy Goat Breeds

The American Dairy Goat Association (ADGA) recognizes nine breeds as dairy goats, based on production, butterfat, and the animal's size.

Full-size breeds of dairy goats include the Saanen, Lamancha, Toggenburg, Alpine, Nubian, and Oberhasli. Nigerian dwarf goats are smaller and yet an excellent producer of high-quality milk. Often a smaller goat breed such as the Nigerian dwarf is exactly what a family will look for when organizing a dairy goat farming business plan.

Saanens originated in Switzerland. They are one of the larger dairy goat breeds. Their milk production is high, and the butterfat content is on the lower end of the goat milk spectrum. Saanen goats are all white or cream colored. The Sable goat breed is related to the Saanen and is the name for colored Saanens.

Nubian goats are a well-known dairy goat. Nubians have gentle personalities and rather loud voices. The breed is characterized by its Roman noses and long droopy ears. The milk is rich in butterfat. The Nubians are sweet tempered, although many goat owners consider the Nubian one of the loudest goats to keep.

Another popular breed of dairy goat is the Lamancha. They appear earless but actually do have small ears. This breed is accepted in any color and is a good dairy goat. They have a distinct appearance that makes them easy to identify.

Toggenburgs are favored by some dairy farmers because they are believed to have a longer lactation period after kidding.

The Alpine goat has a long and interconnected breed history that also includes some breeding with the Oberhasli and Saanen breeds.

Lamancha goats are sturdy, good milkers with great dispositions. The milk is high in butter fat.

Oberhasli goats were originally called Swiss Alpine goats. They have been active in the United States dairy goat world for over a century. Their distinctive appearance includes the dark brown coat color accented by two darker stripes on the face. They are a popular medium-size breed for dairy and often used for pack goats too.

The Daily Milking

Milking the does has to be done regularly or the animal will be in pain and can develop mastitis. The normal practice is to milk every twelve hours. That's twice a day, every day, for the eight to ten months of milk production. The first step includes cleaning the teats and stripping some milk out, before beginning to milk. Caring for any dairy animal is a heavy responsibility to take on.

Goat owners may decide to leave the kids on the doe for the morning milking and then separate them so that the evening milking can be stored and sold. Other farms may decide to let the kids nurse until twelve weeks and then wean. The milking would begin at this point and be carried out twice a day for the rest of the lactation period.

The Dairy Goat Facility

In a small family operation, you may be able to avoid having separate buildings for housing and milking your goats. With a larger commercial operation, the milking is often done in a separate structure. With either plan, cleanliness is the key to success.

The barn will have stalls for the goats. These may be shared, as goats do not like to be alone. Birthing stalls are necessary on a dairy farm because you won't have milk if you don't have does giving birth. Private birthing stalls allow the does to give birth in a quiet environment and bond with the kids. More information and plans for a simple birthing jug can be found on page 164.

Fencing is needed. Rotational grazing practice should be employed, so count on at least two or three separate grazing paddocks or pastures. Depending on your flock size, you may require more pasture areas. Letting one area lie fallow allows regrowth and gives the parasites time to die off. Goats are more likely to escape fencing than sheep, so plan fencing that is strong and cannot be climbed. Goats can jump, too. Be sure the fence is high enough to prevent goats jumping to freedom.

Dairy Options

Are you planning to sell the raw milk to a local dairy for processing? Maybe you are going to produce cheese and yogurt to sell at the farmer's market. No matter what direction you choose to go, having the plan details worked out ahead of time is smart. Contact the proposed buyers of your product and start a business relationship. Each state has different regulations regarding selling dairy products directly to consumers. Check with your local organizations to learn what can be sold from your goat farm including raw milk, cheese, and pasteurized products.

Simple Milking Stand

Photo credit Ann Accetta-Scott.

Milking stands are simple to build and are also useful for grooming and other goat care issues. Having a goat stand will save you much backache, and it will help protect your goat from injury, too. This makes a 36-inch goat stand. The movable 2"x4" moves back to allow the goat to move forward. The board is moved close to the neck and hooked at the top to the fixed head support. Giving the goat some grain to eat while you milk or trim hooves sweetens the time on the stand.

MATERIALS

- 4 pieces of 4"x4" boards, each 16" long
- 11 pieces of 2"x4" boards, each 2½' long (9 for the deck and 2 for the frame)
- 2 pieces of 2"x4" boards, each 36" long (for the frame)

Head Assembly

- 4 pieces of 2"x4" boards, each 2½' long
- 2 pieces of 2"x4" boards, each 4' long
- 1 carriage bolt for head assembly
- 1 hook and eye or small gauge chain (for head assembly, to hold head lock in place)
- Hammer and nails or a drill driver and screws
- Drill and drill bit (for creating a hole for the carriage bolt)

INSTRUCTIONS

1. Build a rectangle from the frame lumber.
2. Insert one 4"x4" leg in each inside corner and secure on both sides of the leg with screws or nails.
3. Turn over onto the legs.
4. Begin adding the deck lumber.
5. Attach the fixed post for the head assembly 12" in from the deck edge.
6. Drill a hole for the carriage bolt in the front 2"x4" of the frame 12" in from the right edge.
7. Attach the chain and hook to the top of the head support board. Screw the eye into the top of the movable head support board.

Milking Equipment

Keep the following items on hand for milking:

- Teat cleaner
- Clean rags or paper towels
- Stainless steel buckets for milk collection
- Quart or half gallon mason jars for storage of milk
- Stainless steel pan or double boiler for pasteurizing milk
- Thermometer to use while heating milk for pasteurizing

Teat Spray #1

By Ann Accetta-Scott, A Farm Girl in the Making

Cleaning the teats before and after milking goats is essential to both udder health and clean milk. Goats lie down wherever they want and often the choice isn't the cleanest spot in the barn. Cleaning the teats ensures that no dirt and pathogens fall into the milking bucket during milking. Washing the teats after milking helps reduce the chance of infection in the udder, referred to as mastitis.

The following teat cleaning formula includes essential oils, castile soap, and colloidal silver.

All essential oils mentioned are gentle enough to apply to the skin and contain antibacterial, antiseptic, and antimicrobial qualities. Lavender essential oil also soothes the skin while providing a calming effect. Castile soap is a gentle natural soap, helping to remove any dirt from the teats prior to milking. Colloidal silver is one of nature's strongest antibacterial, antiseptic, and antimicrobial agents. Colloidal silver can be brewed easily at home or purchased.

The powerhouse of these natural items will not only clean the teats, they help to prevent issues such as mastitis.

INGREDIENTS

- 15 drops lavender essential oil
- 5 drops Melaleuca (tea tree) essential oil
- 10 drops rosemary essential oil

- 3 tablespoons castile soap
- Colloidal silver to fill bottle
- 32-ounce amber or brown glass spray bottle

INSTRUCTIONS

1. Add all contents to spray bottle and gently shake.
2. Spray teats before milking, remembering to wipe the teats prior to milking. After milking, spray and wipe the teats once again.

Teat Spray #2

By Jessica Knowles

INGREDIENTS

- 1 part water
- 1 part rubbing alcohol
- 10 drops tea tree oil

INSTRUCTIONS

1. Add all ingredients to a glass spray bottle and shake gently to mix.
2. Spray teats before milking, remembering to wipe the teats prior to milking. After milking, spray and wipe the teats once again. Note that this formula may have an unpleasant taste for nursing kids. Because the teat cleaner may leave an unpleasant taste, this recipe is recommended for use when kids have been weaned off the doe.

Teat Spray #3

INGREDIENTS

- 20 drops lavender essential oil
- 10 drops tea tree oil
- 2 tablespoons castile soap

INSTRUCTIONS

1. Add all ingredients to a quart-size spray bottle. Fill bottle with warm water and shake to mix.
2. Spray udder and wipe, repeating the process until udder and teats are clean. Proceed with milking. Clean udder with teat spray solution again after you finish milking.

Pasteurizing Fresh Goat Milk

It's hard to find a homesteader or backyard farmer that isn't aware of the amazing health benefits of raw milk, either cow or goat. One fact that we don't hear much about though is making sure that the goat isn't carrying any contagious diseases before we drink the raw milk or use it for cheesemaking.

Having your milking goat tested for disease-causing organisms is an extra step, and it does cost to have the tests run. It's worth it, though, to avoid consuming contaminated milk. The main diseases to test for before ingesting the raw milk are CL (caseous lymphadenitis), CAE, johnes, brucellosis, and Q fever. A positive result will require pasteurization of the milk before drinking or using it in food. CL, brucellosis, and Q fever are transmitted through goat's milk, much the same as cow's milk frequently transmitted tuberculosis decades ago. Refrigerated trucks and knowledge of safer milk handling procedures helped stop the spread of disease associated with milk. Raw milk is considered safe to consume if the animal is clean of diseases that can be transmitted though milk.

Pasteurize your milk if you don't want to test, just to be safe. It is even ok to re-pasteurize the milk received from another source if you want to be completely safe.

How to Pasteurize Raw Milk

When pasteurizing milk, you'll want to be careful to avoid scalding or scorching the milk. Careful monitoring while the milk heats and frequent stirring is required. To avoid scalding, you can use a double boiler, which heats the milk over a pan of hot water. In any event, the stove top heating method is the least risky method.

MATERIALS

- Large pan that can hold the milk
- Thermometer for testing the temperature of the milk
- Ice water bath to chill the milk quickly
- Sterilized containers for milk storage

INSTRUCTIONS

1. Using either the double boiler or just a large pan, heat the milk on the stovetop over medium heat. Your goal temperature is 161°F, and you want it to stay at that temperature for 15 seconds. I use both a long candy-style thermometer and a digital laser style for accurate readings.
2. After the milk is heated and held at the temperature for the 15 seconds, it needs to be cooled quickly. Have a large sink or pan of ice water ready. Put a lid on the pan of milk. Place the pan of milk into the ice water bath.
3. Once the milk has cooled down and won't heat up your refrigerator, pout it into sterilized mason jars or containers and chill completely in the refrigerator before drinking.

CHAPTER 6: Fiber Goats

Fiber goats produce mohair and cashmere. Cashmere goats can be among sixty breeds of goats as they are a type and not a specific breed. Many Cashmere goats are Spanish or Myotonic lines. Goats that are classified as Cashmere fiber goats produce fiber less than 19 microns in diameter, with a length of over 1¼ inch. The fiber should also have a distinct crimp.

Cashmere is harvested by hand, once a year, using a plucking method of collection as the fiber naturally begins to release for the summer weather. Over two hundred grams can be collected from a single goat. The fiber then must be dehaired, usually by hand, which means removing the long goat guard hair. Since mohair and cashmere goats do not produce lanolin like sheep, the scouring process is less intense.

Angora goats produce lovely long locks of mohair fiber. The ringlets are silky and glossy and much sought after in the luxury fiber industry. Angoras are capable of producing over ten pounds of this fiber each year.

Pygora goats also produce mohair. The breed was developed in the 1970s and is smaller than a standard Angora goat, yet bigger than the pygmy goat that was also used to develop the breed. Pygoras produce pounds of soft mohair every six months. Depending on the breeding, the fiber can have little or quite a lot of guard hair to remove. The Type A Pygoras have the most closely related fiber to the Angora breed. The Type B Pygora fiber is a middle of the road combination fiber with some soft long ringlets along with cashmere type fiber. Finally, Type C Pygoras are the cashmere-producing goat. They look to be surrounded by a cloud of soft downy fiber before shearing. Type C Pygoras can be plucked as the fiber begins to release for spring. Type A must be shorn, and Type B, being a combination, may require shearing too. Failure to shear the Pygora goats as you do with Angora leads to felting on the body and uncomfortable skin issues with matted hair pulling on the skin.

Shearing

Part of the fun of owning a fiber-producing goat is shearing day. Angoras, Pygoras, and Cashmere goats are the mohair goat breeds.

Twice a year, usually spring and fall, you will harvest the fiber grown on your fiber goat. This beautiful fiber is a renewable resource on your homestead. The mohair fiber can be curly locks like the Angora goats or more cashmere type fiber. Taking care while shearing the fiber will give you the best product. With Type C Pygora goats, you can choose to pluck the fiber instead of shearing. Plucking the fiber creates a higher quality product because it has less guard hair mixed in. However, you can shear any of the Pygora fiber types.

Preparing to Shear a Goat

Before shearing a Pygora goat, gather clippers, lead and halter, and bags for collecting the fiber. A goat stand, or having someone available to hold the animal, makes the task easier. Stay focused and keep the restrained time as short as possible. Before beginning, pick out any large, noticeable pieces of vegetable matter from the fleece. A blow dryer, with a cool air setting, can help blow out loose bits of hay and bedding.

Since goats are fidgety animals, I strongly suggest using some sort of restraint or goat stand. Restraining the goat's head helps keep it still. You can offer treats as you go to sweeten the process. If you have a goat that tries to kick you the entire time you are shearing, hobbles on the back legs can help.

Before you begin shearing, check the goat over for the location of the teats, wattles, penis, and testicles so you can avoid nicking those sensitive areas.

Pygora goat fiber is classified as three types, A, B, and C. If you are interested in keeping the different types of fiber separate for processing, take note of what type of goat you are ready to shear.

Scissors or Electric Clippers

Spring-loaded scissors, like the Fiskars brand, are a good tool for Pygora goat shearing. When we first began shearing our own goats, we did it using the scissors. It is very time-consuming, however. The benefits of this method include being able to spend some quality time with each goat. They thrive on attention. They love to be loved! As our flock of goats grew, it became necessary to switch to electric clippers. If you only have a couple of fiber goats, you can shear with scissors and enjoy the process, while bonding with your goats.

Using electric clippers, such as the Shearmaster from Oster, finishes the job in much less time. The clippers take some getting used to, and maintaining the clippers is very important. It is easy to make second cuts using the clippers and end up with a lot of short, unusable fiber.

Once your goat is restrained on a stand, you can begin. Take the first swipe with the scissors or clippers. The first cut should be along the top line of the goat, from tail to shoulder area. Use long, smooth swipes with the clippers. Try to complete long, complete swipes first, and train yourself to not go back for a cleanup swipe until the entire goat is sheared. Try to avoid short, choppy clips, which will yield lots of short, unusable fiber. Keep the blades of the clippers or the scissors parallel to the body for the rest of the clip. After the top line is cut, the rest of the cuts should be parallel to the top line, working from top to belly.

Do not pull the fiber away from the goat to shear. This often leads to cuts on the skin. Follow along the lines of the animal's body but do not pull on the skin. After the fiber has been picked up and bagged, clean up any areas that still need to be trimmed.

Gather the clipped fiber and put it in the collection bag, or ask your assistant to gather the fiber while you continue to shear. The neck and chest should be sheared next. Much of this will be good fiber, too. I like to leave the beard on my goats if they have one.

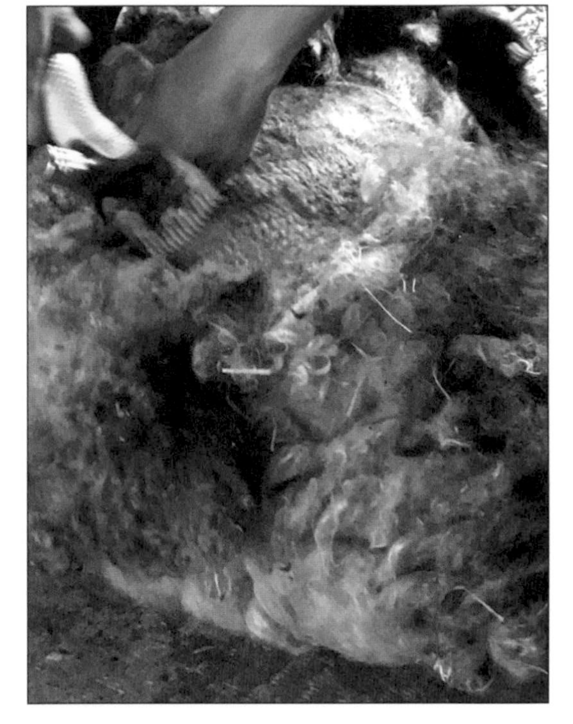

My first attempts at shearing did not have great results. Fortunately, my goats forgave me, and I did improve my skills, over time.

The britch area, belly, and lower legs are not usually processed into roving and yarn. The fiber is often stained, matted, felted, or short in length. Be sure to clip fiber from the armpits, too. Unlike sheep, goats can have armpit hair. Trim off the fiber around the ears and the top of the head. This fiber may be clean but short. It can be saved and used for stuffing things or used for doll hair.

Do not trim the hair at the end of the penis. This hair directs the stream of urine away from the body and is necessary.

It is possible to shear your goats using the same method used by sheep shearers on wool bearing animals. When we have hired professional sheep shearers to shear our goats, most have used this method. It is quick and the fiber is still collected as you would using the scissor method. The shearer starts by clearing the belly, inside lower legs, and britch area of fiber. This undesirable fiber is swept away. Next, the body is shorn, from the belly to the top line, and repeated on the second side. The shearer finishes by clearing the front armpits, the head and neck area, and any bits of fiber that were missed.

A third method that some shearers use on fiber goats is also used for shearing alpacas. The goat shearing is begun while the animal is held in the traditional position. After the belly, britch, and lower legs are shorn, the goat is placed on its side on a mat, and the feet are held in restraints. This lessens the chance of skin cuts because the goat is immobile. It is very quick. One side is sheared, the goat is flipped over, and the other side is completed.

Preparing Mohair for Processing

Once you finish shearing, take the harvested fiber and spread it out on a skirting table. A skirting table is simple to build. Many people construct one using a large wooden frame with welded wire attached to it. The wire allows the short fibers and debris to drop from the fleece. You will still need to pick through the fiber to remove short sections or second cuts, manure tags, felted sections, and vegetable matter. Some fiber goat owners like to keep record of the weight of the usable fiber from each goat.

Raw fiber.

Angora will not require dehairing, but the Cashmere and most Pygora fiber will need to be dehaired. A few fiber mills offer this service, but most do not have the equipment to dehair. Ask specifically about this because while some hair will fall through during a mill processing, not all will. If you require that your mohair be completely free of guard hair, some research will need to be done to find the right mill. The hairs remove easily, so if you only have a couple of animals, it might be far less costly to remove them by hand during the skirting phase.

Collect the fiber in bags, preferably cloth or some breathable fiber, and mark with the date and type of fiber.

Once you have your skirted fiber, you are ready to begin using it for crafts and clothing. Carding or combing the mohair will prepare it for spinning. Learning how to spin wool is a possible next step, although goat fiber can be a challenging roving to start spinning. Many people will blend in some heavier wool fiber to give the roving some more body and strength for spinning. If you are a more experienced spinner, spinning mohair roving can yield a beautiful lace-weight yarn. Mohair fiber also lends itself to felting projects. Mohair dyes easily and is a lasting beautiful fiber for many types of wool crafts.

Skirting Table

MATERIALS

- 2 pieces of 1"x2" boards, each 4' long
- 2 pieces of 1"x2" boards, each 2' long
- 4 corner brackets
- 1 section of 1" welded wire large enough to cover the frame and attach to the sides
- Hammer and finishing nails, or a heavy-duty staple gun

INSTRUCTIONS

1. Build the frame using the corner braces for support.
2. Attach the wire to the wooden frame.
3. To use, set the frame over two sawhorses, or four chairs with the same height chair backs.

Solar Dyeing Mohair

Mohair is easily dyed with both commercial acid dyes, purchased powder extracts of natural dyes, and by using foraged dyes from local plant sources. The fiber can be dyed before spinning or after it is spun into yarn.

Solar dyes with some of your local plants and products found in the kitchen are a simple, cost effective way to dip your feet into dyeing your fiber. Often there are no extra costs involved, and the products are readily at hand.

Hibiscus Solar Dye for Mohair Fiber

This project calls for pre-mordanted fiber or yarn. To pre-mordant, prepare by soaking fiber in a solution of 12% alum per weight of fiber. Bring the alum solution to a simmer and hold for 30 minutes. Avoid agitating the fiber, which can cause felting. Gently squeeze out excess water and carefully transfer to the mason jar of dye.

MATERIALS

- ½ cup organic hibiscus tea petals
- Quart or half gallon mason jar
- Pre-mordanted fiber or yarn
- Cheesecloth
- String
- ½ cup vinegar

INSTRUCTIONS

1. Add ½ cup of organic hibiscus tea petals to a quart or half gallon mason jar of warm tap water. Tie the hibiscus petals in a piece of cheesecloth to help the yarn or fiber stay free of tea bits.
2. Add the prepared fiber. Set the jar in the sun and let the gentle warmth release the color of the tea into the fiber. While solar dyeing is low cost and convenient, it does take several days to reach a deep shade of color.
3. Rinse the fiber once you are happy with the color. Use cool water to rinse and add ½ cup of vinegar to the final rinse to help set the color. Squeeze out the excess water and dry flat away from sunlight.

Onion Skin Solar Dye for Mohair Fiber

Collect onion skins used in your day-to-day cooking. When you have about a cup of packed onion skins available, tie the skins in a square of cheese cloth and submerge in a quart or half gallon mason jar filled with warm water.

Onion skin has natural tannins that will act as the mordant for the fiber. You do not need to pre-mordant the fiber or yarn when using onion skin dye.

Place the jar in the sun for several days. Rinse the fiber in cool water when you are happy with the depth of color.

CHAPTER 7: Raising Goats for Meat

The goat meat market is increasing. Goat meat is the most popular red meat eaten in the world. Ethnically diverse areas are creating a demand in the United States for goat meat for both cultural celebrations and religious traditions. The number of farms producing goats for the meat market has doubled in the last decade and continues to increase.

Kinder goats. Photo by Dillon Irwin.

Goat meat (chevon) doesn't take much getting used to for those of us not accustomed to eating it. Many compare it to venison. The meat is higher in protein and lower in fat than beef. Marinating the meat before using it in various recipes reduces any game taste, particularly from older male goats. Many of the recipes that call for beef can have goat meat easily substituted.

Particularly if you are breeding for dairy goats to continue lactating, you will end up with excess wethers to re-home, sell live to someone else to butcher, or butcher for your own needs. Goats sold for meat are typically young. If the goat is older than three years of age, the meat will be less desirable.

Expect that approximately 50 percent of the live weight will be available for meat. That is after the non-edible parts are removed. The term butchers use for this is hanging weight, and it includes the bones.

Check with your local agriculture extension agent about any inspections that may be required. The laws vary by state and local governments and rural versus suburban goat keeping.

Selecting Stock

The Boer breed is favored among meat goats because of the high muscle yield. However, any breed of goat can certainly become a meat goat. When setting up a dedicated meat goat operation, choose a breed that does well in your area, and has a good yield. Keep in mind that the dressed yield will be between 45 and 50 percent of live weight. The shorter the time that you must feed and care for the meat goat, the higher your return. Ask questions of any breeder about the yield they are currently seeing from their goats. A breeder selling stock should have a clear idea about rate of gain, and time to market weight for their goats.

Hearty Goat Meat Stew

4 to 6 servings

This recipe can be made with goat or any red meat of your choosing.

INGREDIENTS

- 2 pounds goat meat, cut into cubes
- 1 cup sliced celery
- 4 carrots, cut into bite-size pieces
- 4 medium potatoes, cut into chunks (peeled or unpeeled)
- 1 sliced onion
- ½ cup fresh breadcrumbs
- 2 (16-ounce) cans diced tomatoes
- 1 tablespoon salt
- 2 tablespoons sugar

- 3 tablespoons cornstarch
- 3 tablespoons Worcestershire sauce

INSTRUCTIONS

1. Pre-heat oven to 250°F.
2. Combine all ingredients in a large baking dish or Dutch oven. No need to precook or brown the meat.
3. Cover the baking dish tightly with aluminum foil, or place the lid on the Dutch oven.
4. Cook for 6 hours. The long cooking time allows the delicious flavor to develop!

Marinade

This marinade uses tomato sauce as the base. Less tender cuts of meat benefit from the acidic nature of a marinade, as it helps soften the proteins. When using meat from a young tender animal, the marinade will raise the experience to a gourmet level!

INGREDIENTS

- 8 ounces tomato sauce
- 8 ounces beer
- 1 tablespoon Dijon mustard
- 1 teaspoon salt
- ¼ teaspoon pepper
- 1 tablespoon sugar
- 2 cloves of minced garlic or 1 teaspoon garlic power

INSTRUCTIONS

1. Place the ingredients in a zipper-style plastic bag or a large covered bowl. Add the meat and store in the refrigerator. Cook within 24 hours or freeze for longer storage.
2. Marinate meat overnight or early in the day before cooking. Do not save used marinade.

Homemade
Vanilla
2013
Timber Creek
Almond

Chapter 8: Goat Milk Products

Chocolate Almond Goat Milk Ice Cream

Yield: approximately 36 ounces, or 6 six-ounce servings.

INGREDIENTS

- 2 cups whole goat milk, cold
- ½ cup unsweetened cocoa powder
- 1½ cups sugar
- 1 teaspoon almond extract (if you have tree nut allergies, substitute vanilla extract or use ½ teaspoon almond extract and ½ teaspoon vanilla extract)
- 2 cups goat cream (if you can't find goat cream at your market, substitute regular cow milk cream)

INSTRUCTIONS

1. Pour the cold goat milk into a large bowl. Whisk in the cocoa powder, sugar, and extract flavoring of your choice. If you have a blender, you can use that to mix this portion of the recipe.
2. When the dry ingredients are completely blended into the goat milk, add the cream. Fold the cream in using a large wooden spoon until well combined, but do not beat the mixture.

3. Pour the liquid ice cream into serving containers that have lids, or a large glass bowl with a lid. Freeze until solid.

*Other extracts can be substituted. Try mint extract for a chocolatey minty dessert.

Goat Milk Caramel Dessert Topping (*Cajeta*)

This recipe is easily doubled or halved. The dessert topping can be stored for a few weeks in the refrigerator or frozen for longer storage.

INGREDIENTS

Yield: 3 (8-ounce) jars

- 2 quarts goat milk
- 2 cups sugar
- 2 teaspoons vanilla extract
- ¾ teaspoon baking soda (Keeps the milk from boiling over! Don't skip this!)

INSTRUCTIONS

1. Add all ingredients to a heavy bottom pan (6 quart or larger). Heat over low heat to dissolve the sugar and baking soda. Continue heating over medium heat to the simmer point.
2. As the milk reaches boiling it will begin to froth and foam. Turn the heat down so the mixture remains at a simmer. Stir frequently, every 5 minutes or so, for a few hours. I had the caramel sauce simmering while I did other tasks in the kitchen. For a pourable consistency, I cooked the sauce for three hours. A longer cooking time will result in a thicker spoonable sauce and

a darker caramel color. The milk mixture should be turning to a golden brown as the liquid reduces.

3. Once the sauce begins to thicken, keep a close eye on it. If it does get too thick, you can add water a little at a time to thin the sauce. Store the sauce in the refrigerator for up to a month, or freeze for longer storage.

Variations

Adult Only Version
Right at the end of the cooking phase, add ½ ounce of Kahlua liquor to 6 ounces of cajeta.

Slow Cooker
Cajeta can be made in the slow cooker, which reduces the amount of time you need to stir and watch over the pan. It might take 24 hours as the method is much slower, but the result is just as delicious! One thing about cajeta is that you can cook it longer for a thicker sauce or less time for a more pourable result.

Coffee Creamer
For a delicious coffee creamer, cook the sauce until just slightly thickened. Shake before adding to coffee.

Goat Milk Yogurt in an Instant Pot

INGREDIENTS

- 2 quarts of fresh goat milk
- 1 envelope yogurt culture

INSTRUCTIONS

1. If using an instant cooker, such as the popular Instant Pot, heat the milk according to the directions from the instant cooker manufacturer for making yogurt. This kills any pathogens and prepares the milk protein for culture. Typically, the instructions will say to heat the milk to 180–185°F.
2. After the milk has been heated to 180–185 degrees, allow it to cool down to 112°F. Add the culture and allow to sit for two minutes. Stir in the culture.
3. Place the inner pot back in the cooker. Press the yogurt button and adjust the time to the directions on the culture.
4. Refrigerate to chill the yogurt.

Goat Milk Yogurt in a Slow Cooker

From Katie Millhorn, Millhorn Farms Dairy

INGREDIENTS

- 2 quarts fresh goat milk
- 1 envelope yogurt culture

INSTRUCTIONS

1. Turn the slow cooker on low and allow it to preheat while you get the yogurt milk ready.
2. On the stovetop, heat the milk to 185 degrees, then cool to 112 degrees. Add the culture for yogurt. Stir.
3. Pour the milk into two or more sterilized jars. Two quarts or multiple single serving jars will fit in the slow cooker. Put the jars into the slow cooker. Put the lid on the slow cooker.
4. Turn off the slow cooker and unplug it. Wrap the slow cooker in foil and cover tightly with a heavy beach towel. The slow cooker will retain warmth of around 110 degrees while the yogurt forms overnight. The yogurt may take a little longer, in some cases, but the consistency will tell you when it is ready.

Goat Butter

Goat butter is delicious. Since goat milk is naturally homogenized, it is harder to collect the cream from goat milk. Goat owners can slowly collect the small amount of cream that does rise to the top until they have enough to make a batch of butter. Freeze the cream until you have enough to whip up the butter. A purchased cream separator can be used if making butter will be a regular task.

Butter making is easily made with 8 ounces of cream, a glass jar, and a tight-fitting lid. Place the cream in the jar and begin shaking. Shake vigorously for about ten to fifteen minutes. Families often share the shaking, passing the jar along to the next person as they tire. The cream will get thick and then become solid. There will be some buttermilk liquid left in the jar. Save that and use it when baking.

Form the butter into a ball and dip it into ice water. More buttermilk will be released. Squeeze gently. Add salt and mix it in if desired. Form the butter into a rectangle or press into a butter dish.

Basics of Cheesemaking

Especially when keeping dairy goats, excess milk can become a problem. No one wants to waste all that delicious good milk and if you don't keep pigs, your options may be limited on who to feed it to. Making cheese from goat milk is an excellent way to consume the excess milk. Goat cheese can take the form of simple feta-style crumbly cheeses, or solid blocks of aged cheddar. Top your pizza with fresh mozzarella and add goat milk cottage cheese to your salad. The recipes are endless. Cheese making also requires larger amounts of goat milk than, say, soap making. The goal in cheesemaking is to create a tasty cheese while removing all or most of the water, which is the largest component of fresh milk.

Cheese Making Equipment

You'll need the following items before you begin making goat cheese:

- Stainless steel pots
- A long stick thermometer that registers to 220°F
- Cheesecloth
- Long handled spoons
- Long handled spatula or knife for cutting curds
- Rennet (the liquid or tablet used to coagulate the cheese)
- Cheese salt (A fine grain noniodized salt. Substitute sea salt or fine kosher salt)
- Cultures (some cheese recipes call for a specific culture. Cultures react in different ways and learning about the cultures is part of the cheesemaking journey.)
- Calcium chloride, citric acid, and lipase may also be called for in specific recipes for flavor or texture

Simple Farmer Goat Milk Cheese

INGREDIENTS

- ½ gallon goat milk (previously chilled)
- ¼ cup apple cider vinegar
- 1 teaspoon cheese salt

INSTRUCTIONS

1. Warm the goat milk to 195°F. Remove from the heat and add the vinegar. Stir a few times and allow curds to form.
2. Line a colander with cheese cloth. Place the colander over a stainless steel mixing bowl to catch the whey.
3. When you see a lot of curds have formed, pour the curds and whey into the colander and let the whey drain through. Lift the curds using the cheesecloth above the whey and gently squeeze out more whey. The curds should be moist but not dripping.
4. Put the curds into a clean bowl and mix in the salt. Taste and add more salt if desired.
5. Form the cheese into a log. Wrap in plastic wrap and refrigerate for 24 hours to allow the cheese to firm up slightly.
6. Serve with green pepper jelly, as shown, or enjoy plain.

Chevre with Herbes de Provence

Recipe by Connie Meyers

This recipe and the next are generously provided and explained by Connie Meyers. Connie is a well-known speaker and educator on many homesteading topics, including a popular class she teaches on cheesemaking. In addition, Connie has been instrumental in her hometown area for chairing the Northern Colorado Urban Homesteading Tour, a yearly event that highlights and inspires current and future homesteaders. You don't have to have a large plot of land to carry out the traditional homesteading practices of yesteryear. I am grateful to Connie for contributing her knowledge and recipes for this book.

INGREDIENTS

- 1 quart whole milk (use raw if you have it)
- ¼ package direct-set chevre culture (C20G from Cheesemaking.com)
- ½ teaspoon salt (best choices are kosher salt or Himalayan salt)
- 1 tablespoon herbes de Provence

INSTRUCTIONS

1. Allow milk to set out of refrigerator for 30 minutes (so it gradually works its way towards room temperature).
2. Place pot on stove, add milk, and turn unit to medium low heat. Measure out ¼ package of the chevre culture. Sprinkle culture across surface of milk. Allow to set (or bloom) for about 2–3 minutes. Now stir to thoroughly combine culture with milk. (Your stir time should be about 20–30 seconds).
3. Slowly bring milk and culture mixture to 86°F. Take the temperature of the milk frequently so you don't let it get too hot.

4. Once temperature is reached, remove milk from heat. Place lid on pot and allow curd to develop for approximately 10–12 hours. HINT: I start this process after dinner. Then the curd can set up overnight and I can resume cheesemaking in the morning.
5. Line a colander with butter muslin (a much tighter weave than standard cheesecloth). Gently ladle curds into colander and allow to drain about 2 hours. (If I am in a hurry, I give the butter muslin a few gentle squeezes to help remove whey from the curd).
6. Next, set up a small pan with a cooling rack. On top of the rack, place a cheese-draining mat. (Or use another item that is food grade safe with a fine grid, such as a fine mesh strainer. The goal of the fine grid is so the curd does not escape through the grid, but is able to drain). On top of the draining mat, place a cheese mold with weep holes.
7. Spoon the curd from the butter muslin into the mold. Once all of the curd is in the mold, gently press down with the back of a spoon to help remove more whey from the curd. As the curd drains, it will knit together.
8. Allow the curd to drain in the mold for 10–14 hours (or until desired consistency is reached). Note: the goal is to allow the curds time to knit together so they hold the log shape once removed from the mold, as well as enough time to drain so the log is somewhat dry, but not so dry that the log crumbles apart when you slice it.
9. Combine salt and herbes de Provence on a plate or rimmed tray. Roll the chevre log on the herb/salt mixture, gently patting them into place. (HINT: from experience, I can tell you that if the chevre log is too dry, the herbs will not adhere well to the log).
10. Your cheese is now ready to enjoy. You can store the cheese in the refrigerator for up to one week, but it never lasts that long in our house.

TIP: you can skip the store and make your own herbes de Provence.

- 3 tablespoons dried thyme
- 3 tablespoons dried savory
- 2 tablespoons dried parsley
- 2 tablespoons dried marjoram
- 1 tablespoon dried tarragon
- 1 tablespoon dried rosemary, crushed
- 1 tablespoon dried fennel seed
- 1 tablespoon dried oregano
- ½ teaspoon dried lavender (optional)
- ¼ teaspoon dried bay leaf, ground (optional)

You will only use a portion of this herbes de Provence recipe to coat your chevre. The rest may be stored in an airtight container away from direct sunlight until further use.

Goat Milk Cottage Cheese

Recipe by Connie Meyers

MATERIALS

- 1 quart whole goat milk (raw if you can get it)
- 2 tablespoons cultured buttermilk
- 2–4 drops liquid rennet
- 1 tablespoon water
- Salt to taste (do not use iodized)
- ¼ cup cream (optional)

INSTRUCTIONS

1. Place a pot over low heat. Add goat milk to pot. Periodically check the milk temperature, taking care to keep the thermometer off of the bottom of the pot. Heat the milk up to 86°F. Once that temperature is reached, remove the pot from the heat and stir in the buttermilk.

2. Add rennet to the water. Add this to the warm milk and buttermilk mixture. Place lid on pot to help retain the heat. Let the milk set for approximately 1 hour (or until the milk has set a good curd). You'll know it has set when the curd pulls away from the sides of the pot and the whey is a light greenish/ yellow color. If the whey still looks milky, allow it to set a little longer.
3. After the curd has set, slice the curd into approximately ½" squares.
4. Place pot back on low heat. Gently bring the curds up to 110°F, stirring periodically. This stirring helps prevent the curds from matting together. Once temperature is reached, remove pot from heat and place the lid back on the pot. Allow to set for 30 minutes. During this time, the curds will shrink in size, releasing more whey. As the curds shrink, they will become firm(er) in texture.
5. Place a colander over a bowl. Line the colander with cheesecloth. Gently scoop the curds into the cheesecloth. Allow the curds to drain for approximately 20–30 minutes. After this time, dip the curds into a pot of cool water.
6. Allow curds to drain again (this will be a much shorter length of time). Once the curds stop dripping, place into a container and add salt to taste. Add cream if desired.

Goat Milk Mozzarella Cheese

INGREDIENTS AND SUPPLIES

- 1 gallon whole goat milk
- 1½ teaspoons citric acid mixed into 1 cup distilled water
- ¼ teaspoon liquid rennet and ¼ cup distilled water
- ½ to 1 teaspoon salt (according to your taste)
- Cheesecloth
- Rubber gloves
- Thermometer

INSTRUCTIONS

1. Heat milk to 90°F.
2. Add the citric acid mixed in 1 cup cool water. Stir and then heat the milk mixture again to 90°F.
3. Add the rennet mixed with distilled water. Stir briefly.
4. Let the curd form, while continuing to heat to 105°F. Turn off the stove. Remove the pan from the burner. Wait 5 minutes. Watch for the curd to begin pulling away from the side of the pan.
5. Cut the curds using a large knife. Slice into a checkerboard pattern of squares.
6. Using a slotted spoon, transfer the curds to a cheesecloth-lined colander. Place the colander over a large bowl or clean bucket. Continue to form the curd and drain the whey.

7. Transfer to a glass bowl and heat in the microwave for 1 minute. Drain off more whey. Microwave again for 30 seconds.
8. Add the salt and knead it into the cheese. Drain again.
9. Microwave again for 35 seconds. Knead and drain.
10. Knead the curd until it becomes smooth and elastic. When it becomes stretchy like taffy, it is ready!
11. Place the ball of mozzarella into a bowl and place that bowl into a bowl of ice water to chill. Cover the bowl with a light towel or cheesecloth.
12. Grate the mozzarella cheese, if desired. For longer storage, wrap in plastic wrap and refrigerate. Use within three days.

Homemade Pizza Crust

Showcase your goat milk mozzarella on this delicious crust! See below for a sauce recipe, too. Tip: Try using a 14-inch cast iron skillet lightly greased with coconut oil as your pizza pan. Crispy crust and delicious pizza!

Makes 2 (14-inch) crusts

INGREDIENTS

- ¼ cup warm water
- 1 package dry yeast
- 4¼ cups all-purpose flour
- 1 teaspoon salt
- 1 teaspoon sugar
- 1¼ cups warm water
- 2 tablespoons olive oil

INSTRUCTIONS

1. In a large mixing bowl or bowl for stand mixer, sprinkle yeast over ¼ cup warm water. Stir to dissolve. Mix in 2 cups of flour, salt, and sugar.
2. Add the remaining water and oil and beat until smooth.
3. Add the remaining flour 1 cup at a time. Mix and knead until a dough forms.
4. Knead the dough on a floured surface until smooth and elastic.
5. Grease a large bowl.
6. Place the dough in the bowl, turning to grease all sides.
7. Cover with a towel or plastic wrap and place away from cold drafts.
8. Let rise for 1 hour until doubled in size.
9. Punch down the dough. Cut into two pieces. Allow to rest for a few minutes.
10. Using one half of the dough at a time, roll it into a large circle approximately 14 inches in diameter or the size of your pizza pan.
11. Preheat oven 450°F.
12. Top the pizza dough with sauce and cheese and other toppings of your choice.
13. Bake 20 minutes or until crust is crispy at the edges and the toppings are bubbly.

Oven-Roasted Tomato, Garlic, and Onion Pizza Sauce

INGREDIENTS

- 2 cans diced tomatoes
- 6 cloves of garlic minced
- 1 onion chopped
- ¼ cup chopped green pepper, optional
- 2 tablespoons dried oregano (use fresh if you have it)
- 1 teaspoon sea salt
- 3 Tablespoons olive oil

INSTRUCTIONS

1. Preheat oven to 375°F.
2. Add all ingredients to a 13x9 baking dish.
3. Mix gently to combine.
4. Roast for one hour uncovered.
5. Allow to cool to room temperature to avoid splattering hot liquid. Pour the contents into a blender if you prefer smooth sauce for the pizza. Or leave the sauce as is for a more rustic style pizza.

Goat Milk Soap

Goat milk is wonderful for the skin. The lye in cold process soaps changes the molecular structure of the milk, making it shelf stable, while leaving the skin-softening healthy components intact.

When making cold process, lye-based soap, the water and the lye granules are combined, which causes a large increase in heat. The heat is monitored until it reaches the correct level for adding the oils and resuming the process of creating soap.

Milk and lye mixed together will scorch the milk unless you begin with frozen milk. Or you can try a different method for using milk in soap, such as using dry milk powder. I prefer to use the fresh-frozen milk approach. In preparation for making the soap, I measure out the required amount of fresh milk and then freeze it in a Ziploc bag marked with the recipe name. When I am ready to make the soap, I can grab the bag of frozen milk from the freezer and am ready to begin.

Tools for Soapmaking:

- Double boiler or half gallon mason jar
- Large pan for boiling water
- Immersion blender (stick blender)
- Loaf-style soap mold
- Rubber spatula
- Long stick thermometer
- Protective glasses
- Rubber gloves
- Long-sleeved shirt

All tools should be dedicated to soap making and not used interchangeably with food or cooking use.

Traditional Goat Milk Bar Soap

INGREDIENTS

Oils

- 425 grams olive oil (15 ounces)
- 198 grams coconut oil (7 ounces)
- 57 grams castor oil (2 ounces)
- 115 grams sweet almond oil (4 ounces)

Lye Mixture

- 4 ounces frozen goat milk
- 113 grams distilled water
- 111 grams lye granules

INSTRUCTIONS

1. Weigh the solid oils and place in a large glass jar or double boiler. Weigh the liquid oils and have them ready to add to the solid oils when melted. When the solid oils are melted, turn off the heat and mix in the liquid oils. The oil mixture will stay warm sitting in the hot water while you make the lye solution.
2. Wear protective glasses, long sleeves, and rubber gloves when working with lye. Stir carefully and gently without splashing. When using an immersion blender for mixing the soap, always keep the blender fully submerged in the mixture so that it doesn't cause a spray. Turn the

blender off before lifting it from the mixture.

3. Place frozen goat milk in a plastic container (metal bowls will get very hot!).
4. Pour the water over the goat milk.
5. Begin to slowly sprinkle the lye granules over the frozen milk. Do this a little at a time to prevent scorching the milk. Stir gently as the milk begins to thaw. Once all the lye is added, stir until no frozen milk remains.
6. Check the temperatures of both the lye solution and the oils. They should be almost the same temperature.
7. Using a plastic bucket or deep plastic bowl, pour the oils into the container. Have the immersion blender ready, or the spoon if you are stirring by hand.
8. Carefully pour the lye solution into the oils. Stir gently to combine or begin mixing with the immersion blender.
9. Continue to mix until you see the soap begin to thicken (reach trace). Don't go too far! With an immersion blender this will only take a few minutes. Stirring by hand will take longer. Trace is the point where the soap has thickened enough to leave a trail as the spoon is dragged through the mix. It is sometimes described as a light pudding texture.
10. Stop mixing when you reach trace. Add any essential oils, Vitamin E, or fragrance. Pour into the mold. Tap

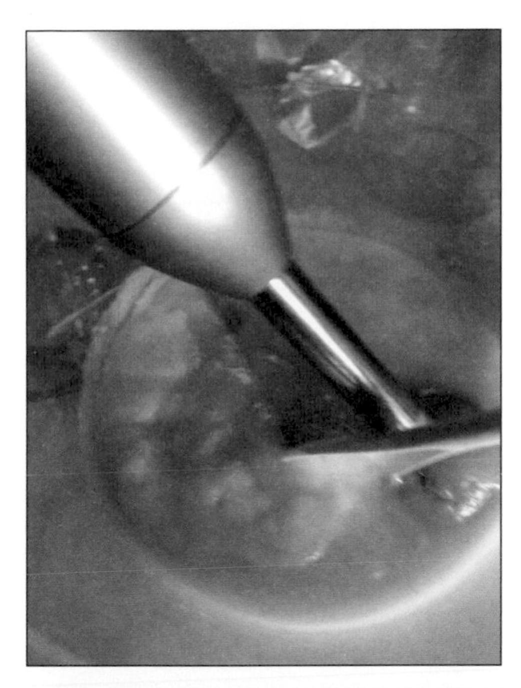

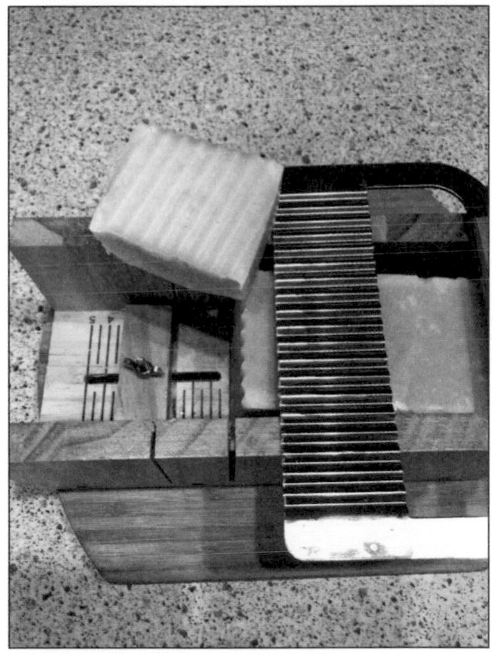

Soap in mold.

lightly on a table to remove air bubbles. Smooth the top to your liking. Some people like it bumpy with ridges, others prefer a smoother surface. If your home is warm, place the soap mold, covered with wax paper, in the refrigerator for 24 hours.

11. Check the next day and, if the soap has formed, remove from the mold and place on a cutting board.
12. Cut the soap loaf into bars. Cutting the soap at this point is not hard to do. As it further hardens, it will become harder to cut. I usually cut the soap loaf by 24 hours unless it is still too soft.
13. Place the bars of soap on a cooling rack lined with parchment paper. Most soap bars take 4 to 6 weeks to completely cure. Curing allows the soap to fully mature and it will last longer than a soft uncured bar.

Goat Milk Shampoo Bars

INGREDIENTS

- 7.5 ounces (213 grams) coconut oil
- 3.5 ounces (87 grams) shea butter
- 10 ounces (295 grams) olive oil
- 3 ounces (85 grams) sweet almond oil
- 4 ounces (113 grams) castor oil (castor oil adds the lather to the shampoo bar)
- 8.5 ounces (241 grams) frozen goat milk (see above notes)
- 0.5 ounces (14 grams) distilled water
- 4 ounces (112 grams) lye granules (sodium hydroxide)

INSTRUCTIONS

1. Have your soap mold clean and ready. Use safety glasses and rubber gloves to avoid any accidental splash burns.
2. Using a double boiler pan or a half gallon mason jar set into a pan of boiling water, begin to melt the coconut oil and the shea butter. When the solid oils are melted, add the liquid oils (except the essential oils) and stir gently to combine. Keep warm.
3. In a large pitcher or measuring cup used only for soap making, add the frozen goat milk and the water. Slowly add small amounts of the lye, about a tablespoon at a time, as the milk melts. Stir gently after each addition. When you add the last of the lye, the mixture should be melted and close to 95°F.
4. When both mixtures are ready, pour the oils and fats into a plastic bucket or bowl as shown in the photos. The chemical reaction creates heat, so using a metal bowl to mix soap can lead to burns.
5. Once the oils and fats are in the bucket, slowly pour the lye and milk mixture into the fats. Stir. Using the immersion blender, mix until the soap is slightly thickened and forming a "trace" when the spatula is pulled through. Trace is the term used in soap making lingo to describe when the soap is ready to be added to the mold. This does not take long to reach when using the immersion blender but can take much longer when you are stirring by hand.
6. If using essential oils to fragrance your shampoo bars, add these at trace. The combined amount of essential oils should stay around 1 ounce or 30 to 33

grams. Stir and then pour the soap into the mold. Cover the mold with wax paper or the cover it might have come with.

7. Place the mold in a cool dark place. If your home is warm, place the mold in the refrigerator for 24 hours. Check after 24 hours. If the soap has set (it might still be sticky), remove it from the mold. Cut the bars, or wait another 24 hours if the soap is too soft to cut. A soap cutting tool is nice but you can use a sharp knife. Place a sheet of paper under the soap with cutting lines marked 1 or 2 inches apart. This will help you cut evenly sized bars.
8. The soap bars will need to cure and complete the drying process. Place them on a wire rack lined with wax paper. Wait four weeks before using. Allowing the soap to completely cure gives you a bar that has completed the process and a soap that will last longer.

Felted Soap Bars

Some people call this "soap in a sweater." The felt makes your soap last longer and provides gentle exfoliation for your skin. Grab your bar of soap and some mohair wool roving or yarn. If you don't have mohair yarn or roving, an all wool yarn will work just fine.

MATERIALS

- Bar of soap
- Mohair wool roving or yarn
- Nylon stocking
- Washboard, bubble wrap, or other textured surface.

INSTRUCTIONS

1. Completely wrap the soap in yarn or fiber.
2. Slide the wrapped bar into a nylon stocking.
3. Using a washboard or other textured surface, rub the soap on all sides while holding it in hot water. The friction along with the hot soapy water will felt the wool or mohair covering the soap.
4. When the felt process is to your liking, squeeze out excess water and soap and let the bar air dry.

Goat Milk Lotion

Goat milk lotion sounds like a creamy soft lotion that would help nourish your tired, dry skin. While that is true, goat milk lotion also has the very high potential to grow harmful bacteria, molds, and other undesirable growth that you do not want to rub into your skin. I chose to not include a goat milk lotion recipe in this book after diving into some research about how goat milk lotion needed to be stored, how and why the preservatives don't always work, and what preservatives are necessary in order to make a shelf-stable, safe goat milk lotion.

To me, the purpose of making my own body products is to eliminate or reduce the amount of chemicals that are added to commercial products for bath and body. I prefer to use Vitamin E liquid to help keep products fresh. But in truth, it is not a preservative. Vitamin E and rosemary can help slow the oxidation of the oils in the lotion, but they're not fail-proof. Some essential oils can also add preservative qualities to a batch of soap or lotion, except when dealing with fresh milk. The milk acts as a medium for the bacterial growth, feeding the bacteria and providing a perfect home for molds to grow. Ick. I sure don't want to rub mold and bacteria laden milk into my skin, and I won't pretend that you should either.

Although I had every intention of making a creamy goat milk lotion as one of the projects in this book, I instead provide a lovely, emollient rich lotion project to soothe your rough hands after a day building goat projects, milking goats, and trimming goat hooves. Store your fresh lotion in the refrigerator for maximum shelf life and a cool, refreshing lotion experience.

After Work Lotion for Barn-Weary Hands

You can add essential oils to customize this lotion. Lavender and chamomile are good for promoting relaxation. Ginger and peppermint are energizing. Try lemon and ginger essential oils for a mid-day pick-me-up aroma.

INGREDIENTS

- 20 grams emulsifying wax (do not substitute beeswax or shea butter)
- 15 grams herb-infused oil (see box on page 156)
- 120 grams distilled water
- Natural preservative (optional)
- Essential oils (optional)

INSTRUCTIONS

1. Using a quart mason jar, add the emulsifying wax and the herbal infused oil. Place in the double boiler or crockpot water bath to melt the wax.
2. In a separate jar, heat the water so that the two liquids are close in temperature.
3. Add the contents of both jars to a mixing bowl. With a small whisk, mix the two liquids together for 30 seconds, just enough to mix. Place the bowl in an ice water bath to speed cooling. Stir occasionally. The lotion will thicken as it cools.
4. When almost room temperature, add any preservatives you choose to add. Also add any essential oils you prefer for fragrance.
5. Store lotion in the refrigerator for a longer shelf life. Most homemade lotions have a recommended 2-week shelf life when refrigerated.

*Note about natural preservatives: Without adding a preservative to water-based lotions, make small batches, store in the refrigerator, and use within two weeks. Natural preservatives are available from suppliers. Natapress, derived from radishes, Leucidal, and Phytocide Aspen Bark are newer preservatives to arrive on the natural components market. Deciding to use a more natural preservative is a personal choice. While the newer preservatives are less chemical laden than conventional products, they still may cause irritation.

Infused Oils—Simple Method

Making infused oil may sound like a complicated involved procedure but it's really very simple to do. There is a long method and a quick method, and both are easy procedures.

Gather the following materials:

1. Pint-size glass jar
2. Approximately one cup of dried herbs of your choice or a combination according to a recipe. I most often prefer chamomile, lavender, calendula, dandelion flower, plantain, thyme, comfrey and wild violet. Experiment with your favorite herbs too. Always use dried herbs when making oil infusions.
3. Olive oil, sweet almond oil, or grapeseed oil.
4. Fine mesh strainer.

With the slow method, place the herbs in the jar, cover the herbs with the oil, filling the jar two-thirds full. Apply the lid. Set the jar in the sun for a few days or in a sunny, warm location in the house.

The quicker method, which I admit to using almost all the time, is to place the jar into a pan of water. Bring the water to a simmer and continue for 20 minutes at a low simmer. Turn off the heat and allow the jar to sit in the warm water for 4 hours or longer.

An alternate way to use the heat method is to place the oil and herbs jar in a few inches of water in a crock pot slow cooker. Keep warm for a few hours as described above.

Strain the oil and herbs, catching the herbs in the fine mesh strainer. The oil should be clear of bits of herbs. Re-strain if necessary.

Continue on with the project calling for infused oil or store covered in the refrigerator.

CHAPTER 9: Breeding Goats

Photo by Carrissa Larsen.

What's that smell? When you have a buck (intact male goat) or you borrow one for breeding season, there's no escaping the perfume in the air. More than any other animal on the farm, the buck will alert you to the beginning of breeding season. In order to keep your milking does producing, breeding will be a fact of life.

As the buck takes notice that the does are going through interesting changes, he too responds with unusual behavior. If you are familiar with goat breeding, you probably join me in wondering how this unique dating arrangement even gets off the ground. The buck will rub his horns on fencing, feeders, or you. The rubbing releases and odor from the scent gland that is very pungent. It's not a scent you will enjoy having on your barn clothes. And if that isn't weird enough, he will start an odd behavior of peeing on his front legs, wipe it on his upper lip, curl back the upper lip, and flap his tongue.

The does have their own unusual behavior when in season. Most breeders keep the does housed separately until the actual breeding dates. This way the delivery date can be tracked more accurately. Since the does don't have a buck readily available to mate them, they often mount each other. Thankfully the does do not have an unpleasant odor associated with their breeding season, although they do become very loud and vocal.

This unpleasant odor will often taint the milk from any does still in milk, so it's a good idea to house the milking doe separately during rutting season. They are calling for a mate, and if you have a buck on the property, you will hear him answer the call with his own distressed yell.

Most goat breeds are seasonal breeders. As the days shorten in late summer and early fall, estrus is triggered. July to December is the usual window of breeding with most goat breeds. The females will become more anxious and unsettled. Does may not eat as much, and bucks may become difficult to handle. The does' reproductive tracts will release eggs every three weeks on average, but often the cycle is shorter. There will be much drama, goat calling, crying, and bleating. The buck will call back, rub on the fence, pace, and generally drive himself crazy trying to get to the ladies. Fencing should be tight and well built! We use a double wooden fence with livestock panels attached to the fence on both sides. The fence is over four feet tall. For some breeds, the fence may need to be higher. Bucks in rut can be very persistent. Many goat breeders have stories of a surprise breeding that can happen when a buck succeeds in breeching the fence line.

Breeding can occur in the male's pen or the doe's pen. My preference was to take the doe to the male for the duration of her heat cycle, then return her to the doe pen. If she did not have her next cycle, I assumed she was pregnant. Make sure you note the date that you took her to the buck so you can prepare the birthing stall in plenty of time. If you house the buck with the does all the time, it is nearly impossible to get an accurate due date. Some of you might be wondering, why not just let nature takes it course? Goats have been breeding in the wild for centuries. While this is true, if you are counting on kids to sell for farm income, or milk from your does, it is good management to know when to expect the kids. Most deliveries will proceed without human intervention. But you may have the opportunity to save kids' and does' lives if you are on hand and informed about when to assist with delivery. This will save you heartache and dollars in the long run. Knowing when the doe was bred lets you know approximately when she will deliver her kids. Gestation length is from 145 to 155 days.

Pre-Breeding Checklist

Evaluate the doe and her gestation history. Take a good look at her body condition. Is she over- or underweight? Any signs of anemia? Does she have any history of congenital illness that could cause problems with her offspring or prevent her from having a healthy pregnancy?

Age of doe—While females are sexually mature at five to six months, it is better to wait until over a year in age to breed the doe.
At the other end of the age question is when to retire an older doe. Is her condition still strong enough to support a healthy pregnancy? Does will continue to cycle for their entire lives. Keep your retired doe separate from the buck or you might risk her life.

Size of doe—If the doe is on the smaller side for the breed, match her with a smaller buck. This will reduce the possibility of a kid that is too large for the doe to birth easily. Should you flush the doe before breeding? Flushing refers to adding grain or high-quality forage to the does' diet to bring up pre-pregnancy weight. Flushing can increase the chances of a doe releasing more eggs during estrus, increasing the chances of a successful breeding.

Doe's gestational history—Did she have trouble with previous births?

Doe's post kidding behavior—Did she reject the kid? How was her milk supply? Did she have mastitis?

None of the above factors should, in themselves, eliminate a doe from your breeding plan. They simply help you to be prepared. A doe that has a bad pregnancy or delivery, or was not a stellar dam the first time, may be a model example of a good mother the next time.

Seasonal Breeders

Most goats, including most dairy breeds (except Nigerian dwarf) are what are referred to as seasonal breeders. They come into estrus for a few months out of the year. Each cycle lasts for approximately twenty days and during two of those days the eggs are released by the ovaries. If breeding occurs during those days, the doe will likely become pregnant.

Non-seasonal breeding goat breeds include:

- Pygmy
- Nigerian Dwarf
- Boer
- Spanish
- Myotonic or fainting breeds

Non-seasonal breeders can cycle year-round and become pregnant during any of the cycles.

Photo by Carrissa Larsen.

CHAPTER 10: Gestation and Kidding

For much of the approximately 150 day gestation, no extra care is required for the doe. If the doe begins pregnancy in good condition, she will likely breeze through. Beginning at around the 100-day mark, increase the feed according to recommendations for gestation. This information is printed on the grain bag. If you normally do not feed grain, increase the amount of grain slowly over many days. In selenium deficient areas, a selenium injection is often advised. You can find out if this is necessary from either your livestock veterinarian or the local agriculture extension officer for your area.

Preparing for labor and delivery includes setting up a small area for the doe to go to for delivery. Take the goat to this stall or birthing jug as her gestation period nears an end. Birthing jugs can be easily set up using some simple materials. The main purpose is to keep her from being bothered by her herd mates and to keep the newborn safe while the mother and kid bond. The size of a birthing jug is small. A 4'x8' birthing jug can be easily built using 6 pallets and a makeshift gate. Hang the water bucket on one of the walls to keep the kid from accidentally drowning. Line the floor with plenty of clean straw.

Goat Birthing Jug

MATERIALS

- 5 to 6 used pallets in good condition (if you have something that can be used as a temporary gate you will only need 5 pallets. Otherwise, the sixth pallet can become the gate.)
- 1 roll of 2' chicken wire.
- Hammer and nails or power drill driver and screws long enough to join the pallets at the corners
- Staple gun for attaching chicken wire to the inside of the jug. This will keep kids from getting stuck in the pallets.

INSTRUCTIONS

1. Begin by clearing an area inside the barn or existing goat stall. Place the kidding jug away from drafts and lots of barn traffic.
2. Stand one pallet on end and join to another pallet at the corner. Add a third pallet to the back wall. Next, add the fourth pallet at the end, forming another corner.
3. Add a fifth pallet to the front, forming the front corner.
4. Using sturdy rope or chains, attach the sixth pallet to the front as a gate that will open for entry to the birthing stall. If you have a small metal gate, that can be used instead.
5. Attach the chicken wire to the lower half of the jug on the inside using the staple gun.

Photo by Michelle Nardozzi.

Labor Signs and Birthing Complications

Early signs of impending delivery include the softening of the tail ligaments and the udder beginning to fill. Keep in mind that these do not necessarily happen any significant time before delivery. A mucus discharge can often be seen at the vulva. The dam becomes restless and often wants none of her herd mates near her. She may lie down and restlessly bite at her side, then get up and eat hay. Normal births happen in a short span of time once the active labor begins. In our case, most deliveries happened while I took a break to get a drink or check something else on the farm. But if a doe required help, I was there to assist.

What should you be looking for as far as signs that labor and delivery are not going smoothly? Once the doe begins to actively push, the kid should be delivered in a few minutes, and definitely within thirty minutes. The beginning view should be of the amniotic sac, followed by one or two small hooves. The front hooves should be first, with the little nose tucked between the front legs. Once the head and shoulders are through the birth canal, the rest of the body should slip right out.

Abnormal birth presentations can hold up the process. Many abnormal presentations will require your assistance in order to have a good outcome. Often with these malpresentations you will need to put on an obstetrical glove and lubricant, push your hand into the birth canal and womb, and reposition the kid. It's much harder than it sounds, and being a smaller person is helpful when dealing with a goat in labor distress. You will feel the contractions and need to work quickly and in sync with the goat's body.

My best advice to you as a new goat breeder is to have an experienced mentor available. When possible, visit a friend's goat farm during kidding season. Learn all you can. Veterinarians can be very useful but not always available. A good goat mentor will help you through some tricky situations without batting an eye. Stay calm and learn from both the successes and hard outcomes.

Goat Kidding Kit

Being prepared is the first step to a successful outcome. While that may not be a famous quote, it is my recipe for success. Unplanned pregnancies in the goat world are fairly common, and there's not a lot you can do when your first hint is a new kid waiting in the stall next to mama. Planned breeding gives you over one hundred days to gather supplies, clean stalls, and have everything ready for labor, delivery, and baby goats. There may still be a loss from a malpresentation, birth defects, or an unexpected accident, but your overall kidding survival rate will be much higher if you are prepared to assist with birth complications.

An easy to carry plastic bin is a good container for the supplies you might need. Experience will also guide you on the items to include in your birthing kit. This list is a good start.

Nitrile exam gloves, lots of dry towels, and a suction bulb top the list. Add a lubricant such as obstetric jelly or olive oil, in case you need to assist with turning a kid from a bad birth position. Iodine is useful for dipping umbilical cord ends. Most of the time, the cord will stretch and separate as the sac is cleaned off and the kid detaches from the uterus. Occasionally, a long cord may need to be clamped off. There are clamps on the market for this purpose, or dental floss can be used to clamp the cord before snipping it free from the placenta.

- Scissors for trimming the umbilical cord
- Flashlights
- Obstetrical lubricant
- Betadine (used for cleaning ewe if you must help with delivery)
- Iodine (for dipping the end of the kid's umbilical cord)
- Old towels (keep the towels clean by storing them in large Ziploc bags)
- Umbilical clamps or dental floss for tying off umbilical cords
- Newborn bulb nasal syringe
- Rope to aid in pulling a lamb that is stuck
- Molasses and treats for the doe to give her some added energy after delivery. A warm drink of water with added molasses can work wonders on an exhausted doe.
- A clean container for collecting colostrum, especially if the doe is not doing well and you must bottle feed

Coats for Baby Goats

Photo by Carrissa Larsen.

Newborn kids may need a little help staying warm during the first few days of life. Many goat keepers use a simple option described below. The photo above shows a goat in a hand knit sweater. I love this idea and hope to make a set of these for our next kidding. However, I may not have time for that, and that is where my husband's old sweaters come in handy. No need to run to the thrift shop, and all I need are a pair of sharp scissors. Here's how to make a quick sweater for a baby goat. Note: for newborn kids, you can try the cuff end of the sleeve. It can make a good neck covering for the kid. The second photo shows older kids who needed a little extra warmth after moving to a colder area. For these I used the upper part of the sleeve.

These quick sweaters won't last forever, but they will get you through a week and help the kid acclimate to the new temperatures. You can also finish off the openings for a sweater that won't be so quick to unravel.

MATERIALS

- 1 sweater (men's large or X large; wool is great, but I have used acrylic sweaters too)
- Sharp scissors that will cut through the sweater fabric

INSTRUCTIONS

1. Cut the portion of the sleeve that will best fit the kid. Measure for the length from the mid-neck area to just before the penis on male kids. Otherwise the sweater will become soaked in urine and cause the goat to be chilled.
2. Lay the sleeve portion flat. Measure down from the neck opening ⅔ of the way and back towards the back two inches. Mark this spot.
3. Cut a hole in the fabric where you marked. A single snip will cause the knit to unravel some, so you don't have to cut the entire circle.
4. Repeat on the other side.
5. Gather the sweater up and stretch open the neck opening wide enough to fit over the kid's head. Pull down over the neck. Carefully fit the front legs through the leg openings and smooth the sweater over the body. If the leg openings are too tight, carefully take another snip in the opening.

Bottle Babies

Photo by Carrissa Larsen.

Occasionally a doe will reject her newborn kid. This is sad to watch, and it is one of the times that we need to intervene for the welfare of the newborn kid. The rejection is difficult to understand, but the primary concern is getting colostrum into the newborn and getting it warm and dry. Having the necessary bottles and knowledge will help you give the rejected kid a good start in life.

Feeding a newborn kid is intense at first. Baby goats eat frequently throughout the day and even require night feeding. Dehydration, scours, and a general failure to thrive can occur, so keen observation is needed when bottle feeding.

On the other hand, one of the cutest things you will ever encounter in farm living is a baby goat drinking from a baby bottle. Those tiny kid goats can really work hard to receive the nutrition they need. There could be many causes for rejection. Some of these can be quickly remedied and the kid then allowed to nurse naturally. Other times, nothing we try will coax a doe into accepting the hungry newborn.

The health of the doe is a possible cause for rejection. If she was a bottle kid herself, that can also play a part. Heredity is another factor, along with the difficulty of the birth and subsequent infections. Healthy, strong does will make better mothers. Was the doe in top condition heading into goat gestation? If a doe is not healthy, she may reject her kid. Especially with a first-time mom, health can make a big difference in her mothering instincts. Offering grain to nibble and warm water sweetened with molasses might bring her around and restore her energy. Then you can try again to get her to accept the kid goat.

Maternal instinct is a strong urge. When a new mom sees her kid, she instinctively takes over the care and protection. The doe will clean it off from delivery and encourage her kid to nurse. Keep records of how your does take on mothering. When you notice that a doe is not as strong as she should be in maternal qualities, that could be a trait passed down in her genetics. It's a good question to ask when purchasing a future breeding doe. If the doe was a bottle baby because her mother refused to care for her, take that information into consideration.

Infections of the teats or udder can cause a doe to kick the kid away. If it hurts her to be nursed, she isn't going to be a willing mom. An infection on one side only may cause her to reject one twin.

There are some tactics you can use to try to get the doe to accept the baby. Take care while trying the following ideas. A kid goat can be harmed and seriously injured by a mother that wants no part of being a mom.

Give the doe some time and space. She might be feeling unwell and unable to take on nursing her kid. Try hand milking and bottle feeding the colostrum. If possible, bottle feed near the doe so she can see the newborn.

After a short break, if the doe is more responsive, try to get the kid latched on to the udder again. This is where a birthing jug can be really useful—it'll keep the other goats from interfering with the bonding.

Apply a drop of vanilla on the doe's upper lip and on the rejected kid's anal opening. The vanilla will disguise any scent that might be sending warning signals to the doe. Refrain from heavy soap or perfume scents when in the barn or handling the kid.

As a last resort, you may need to hobble the doe and let the kid nurse. You may need help accomplishing this if the doe is particularly agitated. It may even take a few days of forced feedings to succeed.

Grafting the rejected kid to another calm, accepting doe sometimes works. Of course, this situation will be different for every flock and might be different from year to year with the same doe. The doe that rejects her kid one year might be an excellent mom the next time she kids.

Some goat keepers use a combination of dam raising and bottle-feeding. This practice preserves the bond between the doe and her kids and the kids continue to reap the health benefits of dam raising,

In our goat breeding days, we had one dam that would not accept her kid. The doe was aggressive toward the kid and, for safety, we brought the kid into our house for the first few days. Once the kid was eating well and strong, we returned him to the barn so he could grow up as a goat. Although we continued to bottle-feed him throughout the day, he often tried to nurse from the other does when their kids were eating.

Keep colostrum in the freezer for such occurrences and use either milk replacer or fresh goat milk from the herd to bottle-feed the rejected kid.

You should try to leave the kid with the herd, even if they are being bottle-fed. The argument for this is that the kid goats learn to eat food, drink water, and nibble hay earlier if left in the herd. Unless there is a serious health concern, this is a good plan. Smaller farms often make sure things are going well for a day or two before returning the rejected kid to the herd. Each goat owner will have to assess the situation and make the best decision based on the facts at that moment. Keep in mind that for normal goat behavior to develop, it is important for the kid to learn from the herd.

Using Milk Replacer When Caring for a Rejected Kid

Powdered milk replacers are available in farm supply stores, and fresh goat milk might be available from another one of your does. Most goat keepers will recommend using fresh cow's milk or fresh goat milk over using a powdered milk replacer. The reason is that the fresh milk results in fewer complications with scours and dehydration. Kids can also be given goat milk from a natural grocery store in your area, although that is a somewhat costly alternative. If you cannot obtain fresh goat milk from your own herd, the most economical option is making your own based on the following recipe. Of course, there are other do-it-yourself–type milk replacer options that are similar being used, too. Find the one with the least side effects and stick with that. Switching formulas can wreak havoc on the kid's digestive tract.

Kid Milk Replacer

Pritchard-style livestock nipples are available for purchase and work well, especially for larger breeds of goats. When raising smaller breeds, the kids' mouths are quite small. We found it better to use regular human baby bottles and enlarge the nipple opening slightly.

While feeding the rejected kid from a bottle, it is important to hold the bottle above their head at an angle. This closely mimics how the kid goat would stand while nursing the doe. It allows milk to bypass the undeveloped rumen and go through to the other three stomachs for digestion and nutrient absorption.

INGREDIENTS

- 1 gallon homogenized whole milk
- 12-ounce can of evaporated milk
- 1 cup buttermilk

INSTRUCTIONS

1. Combine ingredients. Store covered in the refrigerator.
2. Shake gently before filling bottles each time.

How Much Milk Does a Rejected Kid Goat Need?

As a general guideline, be prepared to feed every four hours and gradually increase from 4 to 6 ounces at a feeding to 10 to 12 ounces by the end of the first week. These amounts may be lower for smaller breeds. Bottle feedings continue for three months or more, gradually decreasing in frequency until only one bottle is offered each day. The kid should be increasing intake of hay and grain as it grows.

The following bottle-feeding amounts and recommendations are suggestions that worked for us. The breed of goat, rate of growth, and health of the kid will all play into how much they will require. Never force a kid to eat. An orphaned or rejected kid must eat but shoving the bottle into its mouth while it fights your attempts is not productive.

When bottle feeding isn't going well, try different style nipples, or put a drop of molasses on the nipple.

Day one: (6 feedings/day) 6 ounces, every four hours, of colostrum

Day two: (4 feedings/day) 6 ounces, every 4 hours, of milk and colostrum combined.

Days three to fourteen: Gradually increase amounts in each feeding to 10 to 12 ounces, every 6 hours (4 feedings a day)

Beginning with the third week continue with: 10 to 12 ounces, 4 times a day.

From one month to three months of age: 10 to 12 ounces, 3 times a day.

As you begin to see the kids eating more creep feed and hay, slowly decrease the bottle feedings. Reduce to one bottle feeding of 10–12 ounces per day. Check that growth continues and that forage intake increases. The 3 to 4 month old kid will not want to stop the bottle so you will have to observe and decide when the final bottle is given.

The next month should be decreasing in frequency as you see the kid increasing it's grazing and forage intake. Feed 10 to 12 ounces, 2 times per day.

For the final weaning, 10 to 12 ounces, once a day for 2 weeks.

Observe the kid for growth and forage intake.

Note* Our goats were a small breed and did not take as much per feeding as these recommendations.

Acknowledgments

This book, just as my other books, would never have been completed without my husband, Gary. He is the one who brings the building projects to life. I dream up things that would be useful or fun for the farm and imagine how I would create them. Thankfully, Gary knows a lot more about working with tools and building real things. He graciously corrects my non-builder ways and makes the structures so much better than I would, left to my own devices. I can't thank him enough for being my partner in these books, and in life.

Thanks to my editor, Abigail Gehring, for encouraging me to write this book. Thanks for shepherding me through another writing project.

To my many friends who own goats, have owned goats, or want to someday own goats (Don, I'm talking about you), thanks for inspiring me with your questions, projects, knowledge, and love for all things involving goats. Thanks to Ann, Jess, Carissa, Missy, Ruth, and other friends who were only a text message or phone call away when I needed to talk goats and beg for photos of certain goat items. A huge thanks to Justin Scott who brought the hoop house to life. I am grateful that you gave up a Sunday afternoon to pull this one together for the book.

Acknowledging with a grateful heart my fellow goat owners, photographers, and writers who willingly and gladly shared photos with me for use in this book. The book would not be nearly as rich in breeds and style without your contributions. Michelle Nardozi, Michelle Fraser, Mandi Chamberlain, Elizabeth Taffet, Victoria Young, Denise Smith, Carrissa Larsen, Connie Meyer, Ann Accetta-Scott, Dillon Irwin, Mindie Dittemore, and Lesa Wilke, thank you from the bottom of my heart. Having only a small flock of Pygora goats and two Nigerian goat kids, at this point, my photos would have not told a complete story of the rewarding, and often frustrating, but always heart-warming life with goats.

Thanks to my friend Ruth Lamb for driving me all over her farm so I could get the goat photos and the beautiful cover photo.

Last but certainly not least, thanks to my very talented friend, Jacqui Papi Shreve, for once again providing just the right drawings to round out the book. Your friendship and your talent bless me.

As I always feel at the end of a book writing journey, it only comes together with some group effort, and I am blessed that so many friends would help me make this book the best it can be.

Resources

The following resources have been used throughout the writing of this book. These are my top choices for information on goats and other small livestock, and I recommend them to you.

Books

Nigerian Dwarf Goats 201: Getting Started: How To Choose, Prepare & Care For Your First Goats, Lesa Wilke, 2019, Kindle Publishing

Natural Goat Care, Pat Coleby 2001, 2012 Acres U.S.A.

Temple Grandin's Working with Farm Animals, Temple Grandin, 2017, Storey Publishing

The Joy of Keeping Goats, Laura Childs, 2011, 2017, Skyhorse Publishing

Raising Goats Naturally, Deborah Niemann, 2018, New Society Publishers

The Homesteader's Herbal Companion, Amy K. Fewell, 2018, Lyons Press

Simple and Natural Soapmaking, Jan Berry, 2017, Page Street Publishing

101 Easy Homemade Products for Your Skin, Health and Home, Jan Berry, 2016, Page Street Publishing

Small-Scale Livestock Farming, Carol Ekarius, 1999, Storey Publishing

The Whole Goat, Janet Hurst, 2013, Voyageur Press

The Independent Farmstead, Shawn and Beth Dougherty, 2016, Chelsea Green Publishing

The Accessible Pet, Equine and Livestock Herbal, Katherine A. Drovdahl, MH, CR, DipHIr, CEIT, 2012, published by Katherine A. Drovdahl.

Websites

Maryland Small Ruminant Page, Sheepandgoat.com

Goat Journal Magazine (part of the Countryside Network of Magazines) https://backyardgoats.iamcountryside.com/

Better Hens and Gardens of Bramblestone Farm, Lesa Wilke, betterhensandgardens.com
Feather and Scale Farm, Carrissa Larsen, Featherandscalefarm.com
A Farm Girl in the Making, Ann Accetta-Scott, Afarmgirlinthemaking.com
Fiasco Farm, https://www.fiascofarm.com/
104 Homestead, Jess Knowles, 104homestead.com
Newbury Farms, Michelle Nardozzi, Newburyfarms.com
The Livestock Conservancy, livestockconservancy.org/
Millhorn Farms, Katie Millhorn, rawsomedairy.com/
Urban Overalls, Connie Meyer, UrbanOveralls.net

Instagram Accounts for Goat Lovers

Victoria Young @theyounghomestead
Denise Smith @dewaynesmithfarms
Jessie Irwin @jirwin_farmerchick
Mandi Chamberlain @Wildoakfarm
Elizabeth Taffet @elizabethtaffet for Garden State Goats
Lilith the Goat on facebook https://www.facebook.com/lilithrhegoat/

Index